W0193157

Advances in
TRACER
METHODOLOGY
Volume 4

ADVANCES IN TRACER METHODOLOGY

Collections of papers presented at the
Annual Symposia on Tracer Methodology

Volume 1 (1963):
Papers presented at the Fifth Symposium
and selected papers from the first four Symposia

Volume 2 (1965):
Papers presented at the Sixth, Seventh, and Eighth Symposia

Volume 3 (1966):
Papers presented at the Ninth and Tenth Symposia

Volume 4 (1968):
Papers presented at the Eleventh Symposium

A Publication of the New England Nuclear Corporation

Advances in
TRACER
METHODOLOGY
Volume 4

A collection of papers presented at the
Eleventh Annual Symposium on Tracer Methodology

Edited by
Seymour Rothchild
New England Nuclear Corporation
Boston, Massachusetts

℗ PLENUM PRESS · NEW YORK · 1968

ISBN 978-1-4684-7534-0   ISBN 978-1-4684-7532-6 (eBook)
DOI 10. 1007/978-1-4684-7532-6

Library of Congress Catalog Card Number 62-13475

Plenum Press
A Division of Plenum Publishing Corporation
227 West 17 Street, New York, N. Y. 10011

PREFACE

The Eleventh Symposium on Advances in Tracer Methodology was held in Boston on October 13–14, 1966. The symposium, which was sponsored by the New England Nuclear Corporation, commemorated the tenth anniversary of the Corporation.

The Symposia on Advances in Tracer Methodology, which have been held regularly since 1957, bring together research workers in a variety of disciplines who share an interest in radioactive tracers. The four volumes, this one representing the record of the Eleventh Symposium, provide a compact and readily available source of information—much of which was usually scattered throughout the scientific literature—of interest to biochemists, pharmacologists, endocrinologists, and analysts for whom tracers have become an indispensable tool.

The editor is especially grateful to Dr. Konrad Bloch, Harvard University, and Dr. Waldo E. Cohn, Oak Ridge National Laboratory, for chairing the sessions at the Eleventh Symposium. The cooperation of the speakers is deeply appreciated. In addition, a special bouquet to my secretary, Mrs. Edith Thompson, for her generous assistance in arranging the various symposia and the publication of the papers.

S. R.

Boston, Massachusetts
November 5, 1967

CONTENTS

Applications of Labeled Compounds

INTRODUCTORY REMARKS BY CHAIRMAN

Waldo E. Cohn

Biology Division
Oak Ridge National Laboratory
Oak Ridge, Tennessee

This particular symposium on tracer methodology coincides almost exactly with the twentieth anniversary of the first shipment of radioactive material by the Oak Ridge National Laboratory to the general scientific public. It might be appropiate, therefore, to pause for a glance back to the years when the radioisotope production facilities at Oak Ridge, which underlie so much of biological research in the past two decades, were being developed [1]. Since my concern with radioisotopes, no more than average since then, played a central role in that development, such a glance may give your chairman a bit of "status" in this company [2].

In a very real sense, my involvement in the development of radioisotope production and distribution in the war and immediate postwar period (1943–47) was unplanned. My assignment — for the prosecution of which I engaged the services, in succession, of E. R. Tompkins, J. X. Khym, G. W. Parker, P. C. Tompkins, R. T. Overman, and others — was to investigate the toxicity (radiological) of the highly radioactive products of the fission of uranium in nuclear reactors. As biochemists, we expected the chemists on our part of the Manhattan Project (now the Oak Ridge National Laboratory) to supply us with the separate, individual fission products in carrier-free form. As realists, we soon understood that the chief concern of the chemists was the isolation of plutonium, whereupon we reacted as biochemists have usually reacted: We set out to do the necessary chemistry ourselves. (It should be noted that Tompkins, Tompkins, and Cohn all received the PhD in biochemistry from Berkeley in the same pre-war period.)

Our work on the fission-product problem took the form of developing ion-exchange chromatography, then a rather prim-

1

itive art not far removed from a simple extraction process, into a highly precise separation process. Column chromatography seemed the method of choice for at least two reasons: (1) it should work best at low concentrations (fission products, being born de novo, should exist at submicroscopic levels) and (2) it lent itself to the simplicity required for remote-control operation behind heavy shielding (the initial mixture would contain perhaps a curie of mixed β and γ activities). In "hot labs" constructed rapidly to our specifications [3] (Fig. 1), with periscopes (Fig. 2) and other devices (Fig. 3) for control either at a distance or behind lead or concrete shielding, we effected our first successful separations within 1 year (end of 1944) [4]. Not only were the biologically important Sr and Ba isotopes easily separated from each other, but most of the rare-earth elements were fractionated to degrees that indicated the possibility—soon the certainty—of complete separation from each other [5].

Fig. 1. Dropping "hot" uranium slug into dissolver through roof of "hot cell." Slug was removed from lead container and transferred in a matter of seconds to dissolver inlet. (P. O. Schallert, Oak Ridge, 1946.) U.S. Army photograph.

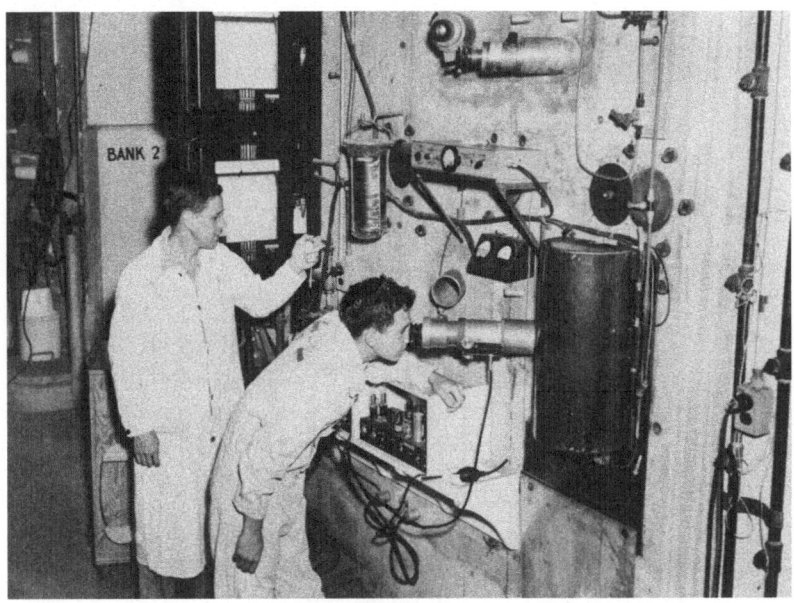

Fig. 2a. Operating face of Oak Ridge "hot cell" during chromatographic separation of fission products (see design in Fig. 2b). Periscope permits indirect line of sight to interior. Lead hemicylinder encloses effluent stream flowing through ionization chamber for the recording of radioactivity. (E.R. Tompkins, left, C. Vanneman, right, Oak Ridge, 1946.) U.S. Army photograph.

The details of this work have been published [3–5] and need not be dealt with here. Significant to the issue at hand is that it provided means of preparing large amounts of radio-isotopes in carrier-free form. A second outcome was the discovery that elution ion-exchange chromatography was a separation tool of unprecedented precision—a discovery that was soon translated, by ourselves and others, into laboratory methods for the separation and/or analysis of nucleic acid derivatives, sugars, amino acids, and the like [6, 7]. Many of the studies reported in this symposium utilize the principles and methods stemming from this development.

Simultaneously with the fission-product work, it became apparent to several of us (particularly those with Berkeley cyclotron experience in pre-war days) that the nuclear reactor might be an effective source of other radioisotopes, particularly

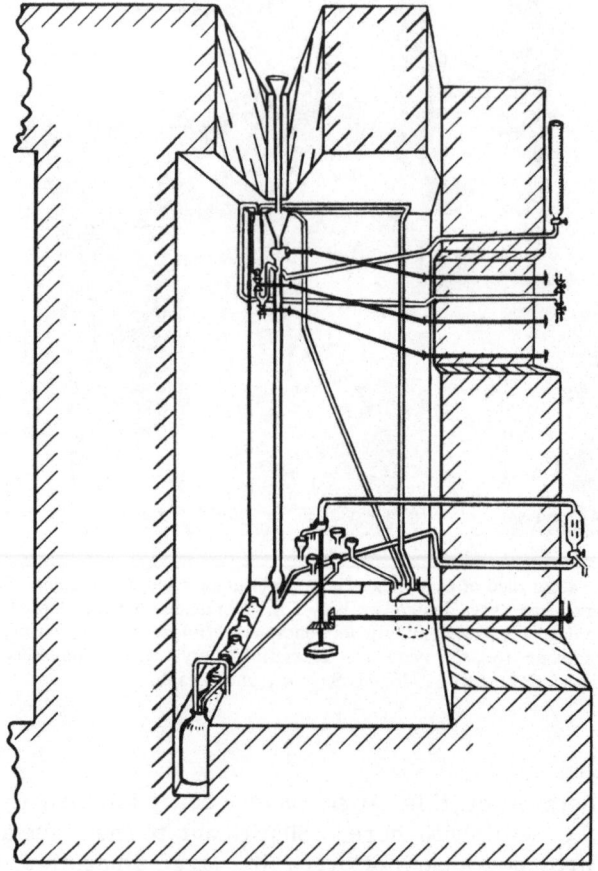

Fig. 2b. Design of remote-control chromatography column [3].

those of greater biological importance, like C^{14}, P^{32}, S^{35}, H^3, and so on [8]. Fortunately, some of these, like C^{14} and P^{32}, could be produced by the transmutation reaction—neutron in, proton out (n,p)—to yield "carrier-free" material of very high specific activity. Accordingly, we began to sneak the appropriate materials into the reactor (after all, the war effort did not, a priori, seem to depend on C^{14} and P^{32} production!). Actually, one of these (P^{32} from sulfur) participated in a cloak-

and-dagger cover up. The Berkeley cyclotron, the principal source of P^{32} for the United States at that time, was completely preoccupied with highly secret uranium experiments for Oak Ridge. So Oak Ridge—namely, myself [9]—was preparing (Figs. 4 and 5) and shipping fractions of curies of carrier-free P^{32} every few weeks to Berkeley to be released from there on a business-as-usual basis, while Berkeley shipped its product, plutonium, to Oak Ridge.

Let it not be imagined that all these developments took place smoothly or in a leisurely fashion. We were engaged in research, yes, but research with materials of unknown characteristics, vanishingly small masses, but infinitely high (for those days) radioactivites; with chemical processes of untested reliability and utilizing relatively novel or unstable structures (the early synthetic ion exchangers, based on a phenol-formaldehyde polymer); and with the potential recipients of our

Fig. 3. Chemical manipulations of radioactive solutions behind lead shielding, with special instruments. (P. C. Tompkins and W. E. Cohn, Oak Ridge, 1946.) U.S. Army photograph.

novel products panting over our shoulders for the material they
needed for their projects. Hot laboratories and remote control
facilities had to be designed and constructed in lead and
concrete on the basis of vaguely conceived chemical procedures
and then put into effective operation almost before the last
construction worker had left. Radioactive contamination had
to be imagined in advance and construction and operation so
arranged as to keep it within presumably safe limits. When
our fingers became contaminated, as happened occasionally and
detergents failed to reduce the amount below that which would
jam a Geiger counter 6 ft away, immersion for 5 to 10 sec in
concentrated HCl often solved the problem (without "solving"
the fingers). A drop of "hot" solution on a concrete floor could
be "hidden" by a lead brick—but when that too was hit, it was
time to scrub both floor and brick. (One wooden floor, under
the P^{32} operation, became so thoroughly contaminated that a
new floor was required.) Glassware with high concentrations
of radioactivity became discolored as the solutions were
evaporated; one early sample of carrier-free radiostrontium

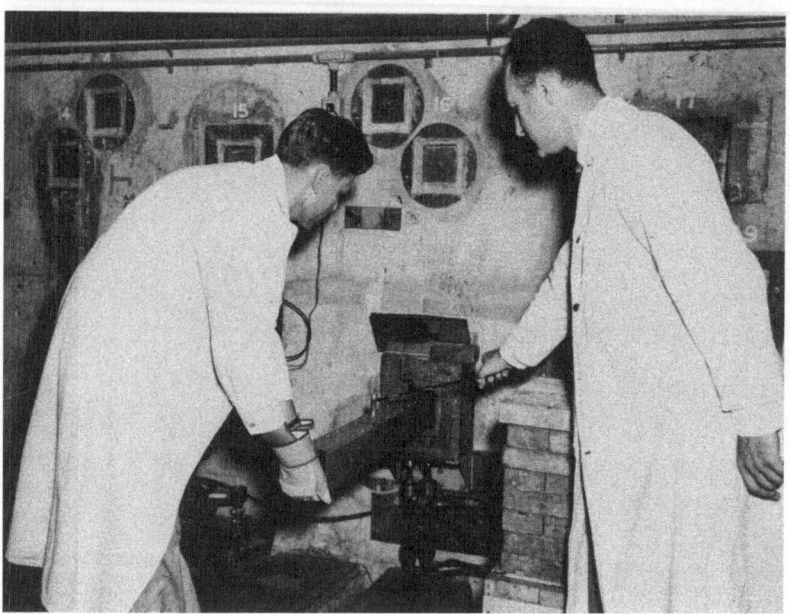

Fig. 4. Placing sample in reactor for neutron irradiation. (Oak Ridge, 1946.)
U.S. Army photograph.

Fig. 5. Removing irradiated samples from reactor. (Oak Ridge, 1946.) U.S. Army photograph.

was laboriously recycled three times to dispose of an iron contamination before it was realized that it was a browning of the glass, not a precipitation of iron salts, that gave the appearance of a solid impurity. A column blocked by the evolution of gas was put back into shape with a long piano wire—but the problem was how to withdraw the wire without exposing the operators to radiation! Some problems were met with shielding, some by distance, and some by speed of handling. Many of the common radioactivity monitors originated with or were adapted to the special hazards encountered in such applied—if then novel—radiochemistry, such as films encased in finger-rings and small, portable direct-reading radioactivity sensing devices. Besides safeguarding ourselves, we had to consider the safety of the reactor—at that time the only one in the world in which samples could be exposed. The rupture of one small can of the wrong kind of material could poison it, or contaminate the area around it beyond reclaiming. Yet the containers

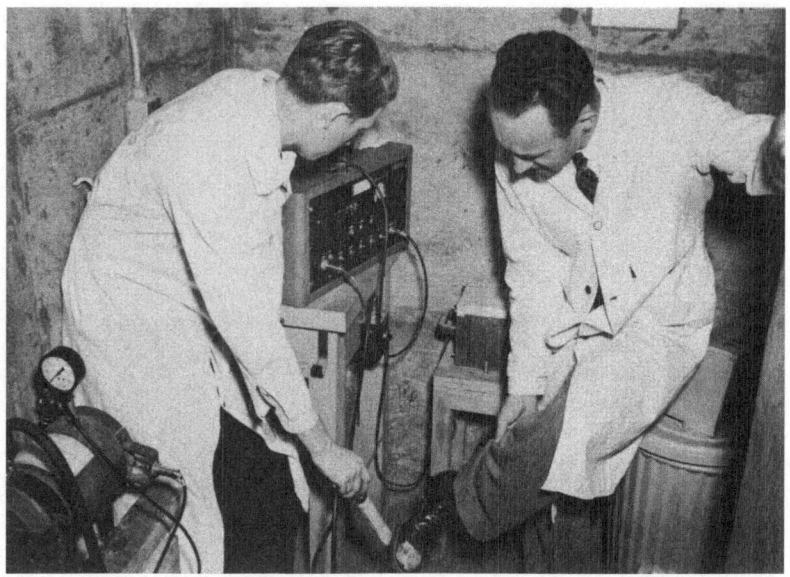

Fig. 6. Checking for radioactivity contamination. (R.T. Overman and W.E. Cohn,
Oak Ridge, 1946.) U.S. Army photograph.

(which could not include more than traces of such neutron
absorbers as boron, for instance, or of potentially strong
γ-emitters, such as cobalt) should be easily opened and emptied
by ourselves or by prospective recipients. Finally, radioactivity
was a word that frightened shipping agencies. Could they be
convinced that properly shielded radioactive substances were
indeed permissible on common carriers? It is perhaps fortunate that we were all young enough to be willing to take on this
complex of problems and, in one way and another, to see them
through.

Both lines of work—the fission product and the neutron
irradiation—were well along by the end of the war, the products
going largely to in-project users. Then came the pressure to
release these to outside experimenters. Some came from
outside, from persons with pre-war radioisotope or radiological
experience, some from inside—as from biochemists like ourselves. Counterpressure, to prevent large-scale production
and distribution, came from project physicists and chemists
who wanted the one-and-only experimental reactor for their
long-bottled-up experiments; from the military-security axis,

who feared loss of secrets about reactor power and also charges
of misuse of government funds expended on a military venture;
and from radiologists concerned with the dangers to personnel
and public. These counterpressures were eventually dissipated,
in large part due to the energetic Paul Aebersold on the
administrative-bureaucratic end, and we were permitted to
publish the catalog we had constructed [10] and to put into use
the production and shipping procedures we had devised at the
prices we had sought to minimize—inasmuch as a true cost
could not possibly be determined.

But once going, the demand proved insatiable, the interference with ongoing experiments manageable, and the costs
less than feared, and the engineers came in as the scientists
who created the program returned to their laboratory exercises.
The first shipment to the outside world took place in August
1946—1 mc of C^{14}—and we all waited anxiously for some time
after that to see if the tenuous thread of production and distribution we had all labored to construct would hold together. It did,
with hasty or improvised repairs here and there as experience
was gained and new problems encountered (such as the presence
of pyrophosphate in phosphate samples), and the widespread
distribution and use of radioisotopes in research and industry
has been a source of satisfaction, over the past two decades, to
all who labored in so many different ways to put it together.
(In 1966, Oak Ridge made more than 5000 shipments, totaling
2.5 million curies.) Thus, though no longer associated uniquely
with either the production or the use of radioactive materials,
I hope I may be permitted some small sense of satisfaction
in the work reported here and in previous symposia dealing
with these substances.

REFERENCES

1. Articles by G.E. Boyd, O.M. Bizzell, A.F. Rupp, and E.E. Beauchamp, in:
   Isotopes and Radiation Technology, Vol. 4, U.S. Atomic Energy Commission,
   Washington, D.C., 1966.
2. Early (pre-war) publications with radioisotopes by W.E. Cohn include: J. Biol.
   Chem. 123:185 (1938), 128:673 (1939), and 130:625, (1939); Proc. Soc. Exp. Biol.
   Med. 41:445 (1939).
3. Cohn, W.E., Parker, G.W., and Tompkins, E.R., Nucleonics 3:22 (1948).
4. Johnson, W.C., Quill, L.L., and Daniels, F., Chem. Eng. News 25:2494 (1947).
5. Tompkins, E.R., Khym, J.X., and Cohn, W.E., J. Am. Chem. Soc. 69:2769 (1947).
6. Cohn, W.E., Ann. N.Y. Acad. Sci. 57:204 (1953).
7. Cohn, W.E., in: Chargaff, E., and Davidson, J.N. (editors), The Nucleic Acids,
   Vol. I, Academic Press, New York, 1965.

8. Kamen, M.D., in: Advances in Tracer Methodology, Vol. 2, Plenum Press, New York, 1965; see also Kamen, M.D., J. Chem. Educ. 40:234 (1963).
9. Cohn, W.E., U.S. Patent No. 2,653,076.
10. Cohn, W.E., Overman, R.T., and P.C. Aebersold (anonymously), Science 103:697 (1946).

RECENT DEVELOPMENTS IN LABELING BY CATALYTIC EXCHANGE—THE APPLICATION OF π-COMPLEX THEORY TO THE PREDICTION OF ISOTOPIC HYDROGEN ORIENTATION IN MOLECULES LABELED BY HETEROGENEOUS TECHNIQUES

G.E. Calf, J.L. Garnett,
and W.A. Sollich-Baumgartner

Department of Physical Chemistry
The University of New South Wales
Kensington, Australia

INTRODUCTION

The value of labeling organic molecules with deuterium and/or tritium in a one-step procedure is now well established for structure determination by mass spectrometry [1], spin decoupling in nuclear magnetic resonance spectroscopy [2], evaluation of coupling constants in electron-spin resonance spectroscopy [3] and in general reaction mechanism studies [4]. Besides specific synthetic chemical methods [5, 6], the most useful general labeling procedures are gas irradiation [7] for tritium or catalytic exchange with isotopic water [8, 9] for deuterium or tritium. The advantages of heterogeneous exchange reactions in the presence of transition metals such as platinum have been critically evaluated when compared with the radiation-induced procedures [10, 11]. A π-complex adsorption mechanism has been proposed to account for the reactivity of molecules in such heterogeneous exchange systems [12, 13].

It is the purpose of the present work to show how the mechanism of isotope incorporation for a variety of organic systems may be rationalized in terms of π-complex adsorbed species as intermediates during exchange. One of the important problems in this type of system is the effect of the substituent on both the total incorporation and orientation of isotope in a

11

given molecule. Recent developments in this field will be summarized using as representative classes of compounds the alkylbenzenes, the polycyclic aromatics, the halogenated benzenes, and the alkylheterocyclics. Particular emphasis will be given to this last series of compounds since these constitute the heterocyclic equivalent of the alkylbenzenes which were most important in the development of the original π-complex mechanisms. For all the work reported in this paper, deuterium will be used as the isotope since this enables the use of mass spectrometry and nuclear magnetic resonance to determine easily the orientation of the isotope without tedious degradation experiments.

EXPERIMENTAL

The general procedure for exchange is to react the organic with isotopic water in the presence of a Group VIII transition metal catalyst in an evacuated sealed tube. The most commonly used catalysts are the activated oxides or chlorides of platinum, palladium, ruthenium, and nickel [14]. Activation is usually achieved with hydrogen [10] or sodium borohydride [15]; however, in specific cases, self-activation (or organic reduction) of the inorganic oxide does possess certain advantages [16, 17].

In a typical procedure, the metallic oxide (100 mg) moistened with water was prereduced with hydrogen or deuterium, the organic (5×10^{-2} M) and heavy water (sufficient to give an equilibrium D of 50%) added, the components subjected to three freeze-thaw cycles, and then vacuum-sealed in a pyrex tube and heated for the required time at the required temperature.

Analytical Methods

For the reactions reported in this manuscript, compounds of high volatility were analyzed on a Metropolitan Vickers MS-2G instrument while those with low vapor pressures were examined on an Atlas CH-4 or Hitachi Perkin Elmer RMU-60, each of which was fitted with a heated inlet. The technique of low-voltage mass spectrometry [18] was utilized, with correction being made for naturally occurring C^{13} and deuterium. Equation

(1) gives the average deuterium content in the molecule ϕ:

$$\phi = \frac{1}{n} \sum_{i=0}^{i=n} i \, di \tag{1}$$

where di is the percentage of aromatic molecules (of n hydrogen atoms) containing i deuterium atoms. The deuterium content was usually in the range 0 to 50% and gave negligible errors from isotope effects associated with the ionization process. Orientation of the isotope in the molecule was determined on a Varian A-60 nuclear magnetic resonance spectrometer.

RESULTS AND DISCUSSION

General Trends in Reactivity

The data in Table I for the exchange of the monohalogenated benzenes and polycyclic aromatics show a broad correlation with ionization potential, particularly for the halobenzenes where benzene > fluorobenzene > chlorobenzene > bromobenzene > iodobenzene [19]. For the polycyclic aromatics a simple relationship between deuteration efficiency and ionization potential or maximum free-valence number [20] is not evident, although it does hold for compounds of similar chemical complexity, e.g., benzene > naphthalene and phenanthrene > anthracene, both groups increasing in reactivity with increasing ionization potential. However, the reverse holds for naphthacene > anthracene > naphthalene.

These results can be rationalized by proposing that the observed trends in exchange rates may be attributed to different strengths of chemisorption and that chemisorption of unsaturated hydrocarbons on a transition metal catalyst occurs through π-complex formation [8, 19]. Reasons for interpreting these reactions in terms of π-complex intermediates in preference to classical associative and dissociative chemisorption have already been discussed in detail [8]. In terms of the charge-transfer adsorption concept [8, 21], a compound possessing the highest ionization potential within a series such as the halobenzenes (e.g., fluorobenzene) will donate π-electrons less

Table I. Exchange Reactions of Monohalogenated and
Polycyclic Aromatics *

Series	Compound	Quantity (moles × 10^2)	D_2O (moles × 10^2)	Exchange time (hr)	Percent approach to Statistical equilibrium	Percent approach to Instantaneous statistical equilibrium	Ionization potential (eV)
A	Benzene[†]	3.0	7.5	3.5	75.0		9.24
	Fluorobenzene	3.0	7.5	3.5	56.0		9.19
B	Fluorobenzene[†]	1.0	2.5	46.5	82.0		9.19
	Chlorobenzene[†]	1.0	2.5	46.5	66.0		9.07
	Bromobenzene	1.0	2.5	46.5	2.0−4.0		8.98
	Iodobenzene	1.0	2.5	46.5	0		8.73
C	Benzene	9.6	17.0	48		100.0	9.24
	Naphthalene	1.25	17.0	48	6.9		8.30
	Benzene	8.0				8.3	
	Naphthalene	2.0	8.0	48	5.1		8.30
	Phenanthrene	1.0	17.0	48	22.0		8.02
	Benzene	8.0				29.0	
	Phenanthrene	1.33	6.6	48	17.4		8.02
	Anthracene	1.0	17.0	48	7.8		7.74
	Benzene	8.0				25.5	
	Chrysene	0.83	17.0	48	23.5		8.04
	Benzene	8.0				92.0	
	Pyrene	1.0	17.0	48	27.8		7.82
	Benzene	8.0				50.0	
	Naphthacene	0.83	17.5	48	11.6		6.64
	Benzene	8.0				78.5	

*All reactions performed at 130°C in presence of Adams catalyst prereduced with
hydrogen (100 mg).
†These compounds were tritiated to specific activities of 10 μc/mg with tritiated
water of the same order of activity.

readily to the catalyst and thus form the least stable π-complex, whereas iodobenzene, with the lowest ionization potential forms the complex of greatest stability.   In this instance of strong chemisorption, specificity may be lost on the catalyst surface, and displacement of the reagent (usually isotopic water) as a source of isotope may occur, leading to autocatalytic poisoning.

For the compounds in Table I reagent displacement effects appear to determine the relative reaction rates. This is demonstrated by the poisoning effect which the polycyclic aromatic hydrocarbons exhibit on benzene in mixed exchange experiments (Table I). Furthermore, the concept of π-complex adsorption may be used to explain why the deuteration of the polycyclic aromatics does not follow a simple ionization potential relationship since the type of bonding in π-complexes depends not only on the ionization potential of the donor and the electron affinity of the acceptor but also on the nature and shape of the orbitals involved in bonding. Thus, polynuclear hydrocarbons with low-ionization potentials and many nodal planes in their bonding orbitals will suffer internal compensation and thus possess a small overlap integral with the acceptor orbitals. Saturated aliphatics possess no π-electrons capable of π-complex adsorption; thus, these compounds are only weakly adsorbed and exchange slowly [8].

Specific π-Complex Mechanisms

Although a molecule may be adsorbed as a π-complex, the actual mechanism whereby isotope is incorporated into the compound needs to be clarified.   On the basis of extensive experimental work [12, 22], two mechanisms have been proposed for exchange between organic compounds and isotopic water. These are the associative and dissociative π-complex substitution mechanisms. In the former mechanism—equation (2)—the π-bonded

$$(2)$$

aromatic is attacked by a chemisorbed deuterium atom originating from the dissociative chemisorption of water. The transition state, except for the π-bonded aromatic, is essentially identical with that of conventional aromatic substitution reactions.

In the dissociative π-complex substitution mechanism —equations (3) and (4)—the π-bonded aromatic reacts with a metal radical in a type of substitution reaction. During this process, the molecule rotates through 90°, resulting in σ-bonded

$$\text{(3)}$$

$$\text{(4)}$$

(edge-on) chemisorption. The postulated transition state for this π-σ conversion occurs at approximately 45° to the catalyst surface and is similar to that proposed by Melander [23] for homogeneous substitution reactions. When σ-bonded, the aromatic undergoes a further but slower substitution reaction with a chemisorbed deuterium atom.

The alkylbenzenes are an important series for the clarification of mechanisms associated with π-complex adsorption. First, this class of compound is much less toxic in mixed exchange reactions with benzene than the polycyclic aromatics [13]; thus, reagent displacement effects do not determine the reaction rate. In the alkylbenzenes, the molecular orbital component in π-complex adsorption remains effectively constant while the ionization potential, electron affinity, and steric factors vary for individual members of the series. Increased methyl substitution decreases the ionization potential [24] but increases the electron affinity [25] and steric hindrance. The first two factors favor π-complex adsorption while the third hinders it. Steric hindrance in different members of the series may be estimated from scale models. Alkyl groups are seen to exercise two different types of steric effects, that is, they may hinder the formation of (1) the π-complex and (2) the σ-bond. The former determines the reactivity of sterically unhindered ring positions while the latter is responsible for the orientation effect. Models clearly show that a *tert* butyl group or two meta-substituted methyl groups prevent the *ortho* carbon from forming a σ-bond with the metal, while such a bond is possible when only one methyl group is involved. In the latter case, steric hindrance is considerably decreased but remains effective enough to produce "severe" deactivation of the ring position.

Exchange data from the alkylbenzenes (Table II, prereduced catalysts) support the above postulates. "Complete" *ortho* deactivation occurs when the *ortho* position is adjacent to a very large inert group or flanked by two methyl groups (*tert*-butylbenzene, benzotrifluoride, mesitylene, m-xylene). "Severe" but incomplete *ortho* deactivation occurs when the *ortho* position is adjacent to only one methyl group. The data in Table II were obtained on prereduced platinum [13]; however, analogous results have been observed with deuterium gas on nickel [26], as well as the exchange between deuterium oxide and the alkylbenzenes over supported nickel [27], borohydride activated nickel [28], and Raney nickel [29].

These *ortho* deactivation effects are important in determining the actual mechanism of exchange, that is, differentiating between the associative and dissociative π-complex substitution mechanisms. The benzotrifluoride results (Table II) show that electronic effects of the substituent do not influence *ortho* deactivation since, although the inductive effect of the $-CF_3$ is opposite to that of an alkyl group, similar *ortho* deactivation to a *tert* butyl group is observed. Furthermore, in homogeneous acid catalyzed exchange of the alkylbenzenes (D_3O^+) no large steric hindrance effects are observed [12, 13]. Thus, steric hindrance in the present work is indicative of the heterogeneous system and would support predominant participation of the dissociative π-complex mechanism rather than the associative which should not give strong *ortho* deactivation. Other more extensive evidence to support this conclusion and also the concept of π-complex adsorption in these reactions has been summarized [22].

Effect of Substituents on Isotope Orientation

Apart from the mechanistic significance of the results of the alkylbenzenes, these compounds are also important because the alkyl substituents are catalytically inert (that is, charge-transfer inactive) in heterogeneous exchange reactions. If the

Table II. Exchange Reactions of Alkylbenzenes with D_2O on Hydrogen Prereduced and Self-Activated Platinum Oxide

Hydrogen prereduced platinum				Platinum self-activated	
Compound	Active aromatic hydrogens	Retarded* reactivity	Ionization potential (eV)	Compound	Percent reaction† completion
Benzene	6	1.0 (120)	9.24	Benzene	40 (130,20)
Ethylbenzene	5	3.0 (120)	8.76	Toluene	53.6 (130,20)
o-Xylene	4	6.1 (120)	8.55	Ethylbenzene	16.1 (130,20)
Isopropylbenzene	3–5	6.7 (120)	8.68	p-Xylene	17.7 (130,48)
Toluene	5	9.3 (120)	8.82	m-Xylene	13.2 (130,48)
Benzotrifluoride	3	13.0 (120)	9.68	o-Xylene	12.6 (130,48)
m-Xylene	3	15.0 (120)	8.56	Isopropylbenzene	3.6 (130,48)
p-Xylene	4	33.0 (120)	8.44	tert-butylbenzene	12.0 (140,48)
tert-butylbenzene	3	350.0 (120)	8.68	Hemimellitene	10.9 (140,48)
Hemimellitene (1,2,3-trimethylbenzene)	3	2400.0 (120)		Mesitylene	9.2 (140,48)
Mesitylene 1,3,5-trimethylbenzene	0	Very high (120)	8.39		

*Measure of relativity of each member of the series under the same catalyst conditions in terms of a rate constant for the active aromatic hydrogens [13], that is, benzene is the most reactive; mesitylene, least reactive. Figures in parentheses are the temperatures of exchange in degrees Celsius.

†Data expressed as percent approach to equilibrium deuterium content for all hydrogens [17]. Figures in parentheses are the temperature in degrees Celsius and the time at temperature in hours.

substituent is no longer inert then an additional charge-transfer interaction is capable of occurring with the catalyst surface leading to changes in the predicted isotope orientation within a molecule. For example, in the monohalogenated benzenes (Table I), fluorine is reasonably inert with the following order of toxicity for the remaining halogens: Cl < Br ≪ I. The monohalogenated benzenes decrease in exchange rate with decreasing ionization potential (Table I), that is, an increase in toxicity of substituent also may be related to a decrease in deuteration; thus, the trend in exchange rates in this series may not be exclusively due to a π-electron interaction but should also be influenced by the decreasing inertness of the halogen substituents which are also capable of charge-transfer adsorption as in species I. This type of interaction

(I)

is supported by the orientation effects in this series. Although severe _ortho_ deactivation is observed for the chloro, bromo, and iodo derivatives, the ratio of ortho/meta deuteration increases from chlorobenzenes to iodobenzenes, that is, with increasing charge-transfer interaction of substituent [19].

Compounds containing other charge-transfer active elements such as nitrogen are also prone to unusual orientation effects. For example, aniline with a lone pair on the nitrogen gives _ortho_-activation effects on platinum and nickel [2, 28, 30] while pyridine (Table IV) and quinoline [2, 28, 30, 31] exhibit α-activation on a variety of catalysts. In these systems, it is clear that an additional charge-transfer interaction involving the lone pair can occur, resulting in the molecule being tilted toward the catalyst surface through the nitrogen. The _ortho_ (aniline) and α (pyridine) positions are thus favorably placed to exchange by the dissociative π-complex substitution mechanism. Thus, it is emphasized that caution should be exercised in predicting (without preliminary experiments) isotope incorporation in compounds containing polar substituents which are likely to be charge-transfer active.

Table III. Orientation of Deuterium in Alkylbenzenes Labeled by Self-Activation *

Run	Compound	Catalyst	Total deuterium (percent)	Deuterium in ring†	Deuterium in α-position†	Deuterium in β-position†	Deuterium in γ-position†
1	Toluene	PtO₂‡	46.8	44.7	50.3		
2	Ethylbenzene	PtO₂‡	45.9	55.0	45.0	31.3	
3	Ethylbenzene	PdO‡	7.0	1.6	14.5	10.7	
4	n-Propylbenzene	PtO₂§	18.2	36.8	9.5	3.0	3.0
5	n-Propylbenzene	PdO¶	11.7	10.0	19.5	7.5	11.7
6	i-Propylbenzene	PtO₂**	25.2	17.6	36.0	29.8	
7	t-Butylbenzene	PtO₂§	6.0	17.4		0.67	
8	o-Xylene	PtO₂**	38.0	22.0	48.5		
9	m-Xylene	PtO₂**	47.3	49.2	45.7		
10	p-Xylene	PtO₂**	28.9	29.5	28.3		
11	Hemimellitene	PtO₂**	29.5	68.3	16.6		
12	Trans-stilbene	PtO₂§	10.4	12.3	1.0		
13	Trans-stilbene	PtO₂**	35.4	42.3	1.5		
14	Tetralin	PtO₂**	40.2	55.5	59.4	5.75	

*Estimated by NMR.
†With respect to total hydrogen at each position.
‡Performed at 130° for 100 hr; reagent quantities as in Table IV.
§Performed at 140° for 48 hr; reagent quantities as in Table IV.
¶Performed at 130° for 48 hr; reagent quantities as in Table IV.
**Performed at 180° for 48 hr; reagent quantities as in Table IV.

An additional problem affecting isotope orientation is the nature of catalyst activation. If self-activated PtO_2 is used (Table II) instead of prereduced oxide, possible changes in isotope orientation may occur because of the differences in mechanism in exchange between the two catalysts [17]. The significant difference is the increase in side-chain/ring exchange on self-activated catalysts (Table III) rather than differences in o, m, p ratios. Self-activated catalysts undergo extensive hydrogen abstraction reactions, particularly in the initial process of catalyst activation where hydrogen for reduction of the inorganic oxide originates from the organic compound. Self-activated catalysts thus possess large amounts of oxygen incorporated into the bulk of the catalyst, and this appears to affect the specificity of aromatic/aliphatic exchange [17].

Alkylheterocyclics

In terms of mechanistic studies, the alkylheterocyclics are important for the prediction of isotope incorporation since this series constitutes the heterocyclic equivalent of the alkylbenzenes. There is now competition between the enhanced adsorption effect of the lone pair from the heterocyclic nitrogen and steric hindrance from the alkyl groups. It is interesting to observe which of the two effects predominates, particularly in compounds such as α-picoline (II) and 2,4-lutidine (III).

(II)   (III)

In Tables IV and V, the results of exchange reactions are reported on a variety of Group VIII transition metal catalysts with 2-, 3- and 4-picolines and 2,4-, 2,5- and 2,6-lutidines. Pyridine (Table IV) has been used as a reference standard for all runs. It is observed that for pyridine, exchange on cobalt activated by borohydride reduction of the chloride salt leads to exclusive α-deuteration. The same specificity has been reported for nickel-on-kieselguhr [31]; however, the cobalt

Table IV. Deuteration of Pyridine on Group VIII Transition Metal Catalysts

Method of catalyst activation	Compound used for catalyst preparation	Deuterium content (atoms percent deuterium)	Orientation; average number of deuterium atoms in molecule in			Deuterium distribution					
			$2-+6-(\alpha)$	$3-+5-(\beta)$	$4-(\gamma)$	d_0	d_1	d_2	d_3	d_4	d_5
Deuterium reduction at 25°C	H_2PtCl_6	7.2	0.35	0.00	0.01	73.2	19.2	6.3	0.9	0.4	
	Nickel-on-Kieselguhr[31]	39.0 ($D_\infty = 92.1$)	1.81	0.07	0.00	0.0	8.5	87.2	4.3		
	$PtO_2 \cdot xH_2O$	43.1	0.90	0.85	0.42	7.3	21.7	33.0	25.9	10.5	1.6
Sodium borohydride reduction at 95°C	$CoCl_2$	22.3	1.09	0.04	0.00	20.2	48.5	31.2	0.1		
	$NiCl_2$	20.1	0.78	0.25	0.02	32.7	39.9	22.0	4.9	0.5	
	$RhCl_3$	32.8	0.86	0.46	0.32	15.1	31.7	32.2	16.3	4.3	0.4
	H_2PtCl_6	27.4	1.06	0.17	0.14	17.8	39.8	31.7	8.9	1.6	0.2

*The statistical equilibrium for all exchanges is 50 atoms percent deuterium. The quantities of reagents used were pyridine, 4.0×10^{-2} moles; D_2O, 10.0×10^{-2} moles; catalyst equivalent to 100 mg of metal. All reactions were carried out without shaking at 130° for 48 hr.

catalyst is cheaper and easier to prepare and the method is thus to be preferred. Chloroplatinic acid is also reasonably specific in catalyzing exchange in the α-position with both hydrogen and borohydride reduction, although the more reactive borohydride catalyst also gives some β and γ deuteration. Platinum oxide catalyzes exchange in all positions in pyridine, although the rate is fastest for the α-hydrogens [8]. Confirmation of the nmr orientation is shown by the low-voltage mass spectrometry deuterium distribution patterns, for example, with cobalt there is a distinct cut-off at the d_2 peak corresponding to the two α-hydrogens. This result would suggest that pyridine is bonded predominantly through its lone pair to the cobalt catalyst (species IV), similar to what was previously proposed for the nickel systems [30–33], whereas on platinum exchange in all positions of the molecule indicates that pyridine is probably tilted during adsorption (species V) as previously proposed [32].

(IV)   (V)

In the picoline and lutidine series, the specificity of cobalt is still retained in the α-positions. If the two α-positions are blocked by methyl substituents (that is, 2,6-lutidine), exchange is confined to the methyl groups only. This catalyst is thus useful for specific labeling in these compounds.

On the borohydride-reduced platinum catalysts, there is a small degree of deactivation of the positions adjacent to the methyl groups in the picolines; however, this *ortho* deactivation is not so severe as with the corresponding alkylbenzenes. This is particularly evident with the 2-position in 3-picoline. A similar trend in reactivity is observed with the lutidines; however, it is to be noted that the 3-position in 2,4-lutidine(III) is completely deactivated in an analogous manner to positions flanked by two methyl groups in the alkylbenzenes (m-xylene and mesitylene). This is to be expected since in 2,4-lutidine the methyl group in the 2-position sterically blocks the effect of the nitrogen lone pair on the adsorption characteristics of the 3-position. The results are thus consistent with the original

Table V. Catalytic Deuteration of Picolines and Lutidines

Compound	Method of catalyst activation	Compound used for catalyst preparation	Deuterium content (atoms percent deuterium)	Orientation average number of deuterium atoms in molecules in				Deuterium distribution							
								d_0	d_1	d_2	d_3	d_4	d_5	d_6	d_7
2-picoline				6-(α)	3-+5-(β)	4-(γ)	2-methyl								
	Sodium borohydride Reduction at 70°C	CoCl₂	10.6	0.79	0.01	0.00	0.05	40.1	47.0	11.7	1.1	0.1			
		PtO₂·xH₂O	28.8	0.60	0.96	0.31	0.16	9.4	24.9	33.6	21.7	8.0	2.0	0.4	
	Self-activation at 130°C	PdO	2.0					87.2	11.8	0.9	0.1				
		PtO₂·xH₂O	15.1	0.14	0.16	0.14	0.63	41.5	26.1	18.6	12.9	0.8	0.1		
3-picoline				2-+6-(α)	5-(β)	4-(γ)	3-methyl								
	Sodium borohydride Reduction at 70°C	CoCl₂	14.4	0.97	0.00	0.02	0.03	25.8	48.4	25.1	0.5	0.2			
		PtO₂·xH₂O	27.5	0.95	0.43	0.45	0.05	9.2	26.5	35.1	22.1	6.1	0.7	0.2	
	Self-activation at 130°C	PdO	1.2					95.3	2.2	1.5	0.7	0.3			
		PtO₂·xH₂O	4.0	0.06	0.00	0.03	0.21	84.2	8.4	3.7	3.1	0.5	0.1		
4-picoline				2-+6-(α)	3-+5-(β)	4-methyl									
	Sodium borohydride Reduction at 70°C	CoCl₂	17.7	1.07	0.03	0.21		21.2	42.5	29.7	5.9	0.5	0.1		
		PtO₂·xH₂O	24.7	1.15	0.40	0.18		10.7	31.9	36.3	16.8	3.7	0.5	0.1	
	Self-activation at 130°C	PdO	5.8	0.02	0.01	0.37		65.9	28.7	4.7	0.5	0.1	0.1		
		PtO₂·xH₂O	37.6	0.93	0.00	1.70		6.0	12.7	24.4	31.3	20.4	5.0	0.2	

2,4-lutidine

			6-(α)	3-+5-(β)	2-methyl	4-methyl								
Sodium boro-hydride	CoCl₂	8.8	0.56	0.04	0.05	0.14	37.8	47.5	12.7	1.8	0.2			
	NiCl₂	13.7	0.43	0.10	0.54	0.15	28.0	36.3	24.1	8.9	2.2	0.4	0.1	
Reduction at 70°C	PtO₂·xH₂O	7.2	0.43	0.10	0.09	0.03	51.6	36.4	9.2	2.1	0.6	0.1		
	PdO	3.3					75.1	21.1	3.4	0.3	0.1			
Self-activation at 130°C	PtO₂·xH₂O	20.0	0.14	0.07	0.81	0.78	28.4	24.5	17.0	12.1	9.0	6.1	2.6	0.3

2,5-lutidine

			6-(α)	3-(β)	4-(γ)	2-methyl	5-methyl								
Sodium boro-hydride	CoCl₂	5.7	0.47	0.02	0.02	0.00	0.00	53.2	43.3	3.0	0.4	0.1			
Reduction at 70°C	PtO₂·xH₂O	24.5	0.42	0.40	0.39	0.02	0.00	14.9	21.6	23.7	20.7	12.5	5.0	1.3	0.3
Self-activation at 130°C	PtO₂·xH₂O	19.2	0.37	0.14	0.28	0.86	0.08	23.6	26.0	22.2	15.7	8.3	3.2	0.8	0.2

2,6-lutidine

			3-+5-(β)	4-(γ)	2+6-methyl						
Sodium boro-hydride Reduction at 70°C	PtO₂·xH₂O	17.8	0.71	0.46	0.43	32.8	22.3	19.9	13.7	5.2	2.9
Self-activation at 130°C	PtO₂·xH₂O	12.4	0.27	0.57	0.27	31.6	40.5	17.6	6.7	2.2	0.7

*The statistical equilibrium for the picoline exchanges is 41.7 atoms percent deuterium. The quantities of reagents used in exchanges were: picoline, 4.0 x 10⁻² moles; D₂O, 10.0 x 10⁻² moles; catalyst equivalent to 100 mg metal.

The statistical equilibrium for the lutidine exchanges is 35.7 atoms percent deuterium. The quantities of reagents used in exchange were: lutidine, 4.0 x 10⁻² moles; D₂O, 10.0 x 10⁻² moles; catalyst equivalent to 100 mg metal.

All reactions were carried out at 130°C for 48 hr without shaking.

alkylbenzene exchange work. From poisoning studies performed on the heterocyclics such as pyridine, the quinolines, and acridine [33], together with the present results from the alkyl-heterocyclics, it would appear that π-complex mechanisms already proposed on the basis of the exchange properties of other systems are also satisfactory for explaining labeling patterns in the alkylheterocyclics. The mechanisms thus would appear to be of general applicability in isotope exchange reactions.

CONCLUSION

Since the early pioneering work with substances essentially of biological importance [34–36], heterogeneous catalytic exchange has been developed as a general labeling tool for deuterium and tritium [8]. The π-complex adsorption mechanism satisfactorily explains the exchange properties of a wide variety of organics. Two important features associated with this concept are steric hindrance effects from bulky substituents and additional charge-transfer interactions with the catalyst surface from substituents which may be charge-transfer active.

ACKNOWLEDGMENT

The authors thank the Australian Research Grants Committee, the Australian Institute of Nuclear Science and Engineering (Mr. E. A. Palmer) for assistance in the purchase of the heavy water, and Commander J. Mason for instrumentation advice. Acknowledgment is also made to the donors of the Petroleum Research Fund, administered by the American Chemical Society for the support of this research.

REFERENCES

1. McLafferty, F.W., Mass Spectrometry of Organic Ions, Academic Press, London. 1963.
2. Garnett, J.L., Henderson, L.J., Sollich, W.A., and Tiers, G.V.D., Tetrahedron Letters 516 (1961).
3. Aylward, G.H., Garnett, J.L., and Sharp, H., Chem. Comm. 137 (1966).
4. Burr, J.G., Tritium in the Physical and Biological Sciences, Vol. I, IAEA, Vienna, 1962.
5. Shatenshtein, A.I., Isotopic Exchange and the Replacement of Hydrogen in Organic Compounds, Plenum Press, New York, 1962.

6. Rothchild, S., Advances in Tracer Methodology, Vol. I, Plenum Press, New York, 1963.
7. Wilzbach, K. E., J. Am. Chem. Soc. 79:1013 (1957).
8. Garnett, J. L., Henderson, L. J., and Sollich, W. A., Tritium in the Physical and Biological Sciences, Vol. II, IAEA, Vienna, 1962.
9. Yavorsky, P. M., and Gorin, E., J. Am. Chem. Soc. 84:1071 (1962).
10. Garnett, J. L., Nucleonics 20:86 (1962).
11. Long, M. A., Odell, A. L., and Thorp, J. M., Radiochim. Acta 1:174 (1963).
12. Garnett, J. L., and Sollich, W. A., J. Catalysis 2:350 (1963).
13. Garnett, J. L., and Sollich-Baumgartner, W. A., J. Phys. Chem. 69:1850 (1965).
14. Garnett, J. L., and Sollich, W. A., Australian J. Chem. 18:1003 (1965).
15. Calf, G. E., and Garnett, J. L., J. Phys. Chem. 68:3887 (1964).
16. Garnett, J. L., and Sollich, W. A., Australian J. Chem. 18:993 (1965).
17. Fisher, B. D., and Garnett, J. L., Australian J. Chem. 19:2299 (1966).
18. Garnett, J. L., and Sollich, W. A., J. Catalysis 2:339 (1963).
19. Garnett, J. L., and Sollich, W. A., Australian J. Chem. 14:441 (1961).
20. Coulson, C. A., J. Chem. Soc. 1435 (1955).
21. Mulliken, R. S., J. Am. Chem. Soc. 74:811 (1952).
22. Garnett, J. L., and Sollich-Baumgartner, W. A., Advan. Catalysis 16:95 (1966).
23. Melander, L., Transition State, Spec. Publ. Chem. Soc. (London) 16 (1962).
24. Watanabe, K., Nakayoma, T., and Mottl, J., A Final Report on Ionization Potentials of Molecules by a Photoionization Method. Contrib. No. DA-04-200-ORD.480 and 737, University of Hawaii, Honolulu, 1959.
25. Matsen, F. A., J. Chem. Phys. 24:602 (1956).
26. Crawford, E., and Kemball, C., Trans. Faraday Soc. 58:2452 (1962).
27. Macdonald, C. G., and Shannon, J. S., Australian J. Chem. 18:1009 (1965).
28. Calf, G. E., and Garnett, J. L., unpublished data.
29. Garnett, J. L., and Sollich, W. A., Nature 201:902 (1964).
30. Lauer, W. M., and Errede, L. A., J. Am. Chem. Soc. 76:5162 (1954).
31. Macdonald, C. G., and Shannon, J. S., Tetrahedron Letters 3351 (1964).
32. Garnett, J. L., and Sollich, W. A., Australian J. Chem. 15:56 (1964).
33. Ashby, R. A., and Garnett, J. L., Australian J. Chem. 16:549 (1963).
34. Bloch, K., and Rittenberg, D., J. Biol. Chem. 149:505 (1943).
35. Fukushima, D. K., and Gallagher, T. F., J. Biol. Chem. 198:861 (1952).
36. Eidinoff, M. L., and Knoll, J. E., J. Am. Chem. Soc. 75:1992 (1953).

THE PREPARATION OF N-METHYL-C^{14}-LABELED DRUGS

Robert E. McMahon
and Frederick J. Marshall

Lilly Research Laboratories
Indianapolis, Indiana

In modern pharmaceutical research practice an ever-increasing emphasis is being placed on studies of drug dynamics, that is, the absorption, distribution, metabolism, and elimination of an administered drug. Radiocarbon labeling can be one of the most valuable tools for such studies, provided that the drug to be studied can be synthesized containing the radiocarbon label. Fortunately, for one very large class of organic compounds of physiological interest, the N-methylamines, a wide variety of labeling procedures are available. Indeed, among the earliest of radiocarbon-labeled drugs to be prepared were meperidine-N-methyl-C^{14} (1), dimethylamino-azobenzene-N-methyl-C^{14} (2), and morphine-N-methyl-C^{14} (3).

The literature of classical organic chemistry presents an abundance of material on the preparation of N-methylamines. This circumstance, together with the fact that there are now commercially available a large variety of suitable radiocarbon-labeled starting materials (see Table I), has resulted in the synthesis of a large number of N-methyl-labeled drugs and other compounds of biological interest. In the following paragraphs the current status of this important area of radiosynthesis is discussed.

1. METHODS FOR THE PREPARATION OF N-METHYL-C^{14} TERTIARY AMINES

Reductive Alkylation with Radioformaldehyde

Probably the most widely used procedure for preparing N-methyl-labeled tertiary amines is the reductive alkylation

Table I. Commercially Available Radiocarbon-
Labeled Intermediates

1. Dimethylamine	7. Methyl chloride
2. Dimethyl sulfate	8. Methyl iodide
3. Ethyl formate (carboxyl-C^{14})	9. Paraformaldehyde
4. Formaldehyde	10. Phosgene
5. Methylamine	11. Sodium formate
6. Methyl bromide	

of secondary amines with radioformaldehyde by the Clark-Eschweiler procedure:

$$R_2NH + H_2C^*{=}O \xrightarrow[\Delta]{HCOOH} R_2NC^*H_3$$

Tarpey et al. [1] have demonstrated that in this reaction it is the carbon atom from formaldehyde and not from formic acid which is incorporated into the N-methyl group. Compounds which have been labeled by this procedure include meperidene [1], (−)nicotine [4], chlorpromazine [5, 6], morphine [7], and N-methyl-4-phenylpiperidine (unpublished, these laboratories). The Clark-Eschweiler reaction cannot be used successfully for the preparation of secondary amines by methylation of a primary amine since the reaction does not stop at this intermediate stage.

In certain cases a metal catalyst plus hydrogen has proved very satisfactory as a substitute for formic acid as a reducing agent. Palladium (5% on carbon) has been used successfully for the preparation of erythromycin [8], propoxyphene [9], acetylmethadol [10], and dimethylaminoethylbenzene [11]. Levorphanol-N-methyl-C^{14} has been prepared by Woods et al. [12] using Raney nickel as catalyst. This procedure is, of course, unsatisfactory if the molecule contains groups subject to either catalytic reduction or hydrogenolysis.

One difficulty which arises when radioformaldehyde is used as a reagent should be mentioned. In these laboratories, as well as in others, it has been observed that 15 to 30% of the radiocarbon present in commercial radioformaldehyde is not present as formaldehyde. This impurity does not appear to be a polymeric form of formaldehyde since heating in water at 110° in a sealed tube does not increase the amount of available formaldehyde—nor is the impurity radiomethanol which is present only to the extent of 1 to 2%.

In any event, the presence of this radioactive impurity does not appear to interfere with the methylation reaction. Radio-carbon-labeled paraformaldehyde, of course, is, available and is readily and quantitatively converted to radioformaldehyde by heating in water at 110° in a sealed tube. The additional expense involved in the use of paraformaldehyde-C^{14}, however, is usually not justified unless very pure radioformaldehyde is required.

Alkylation with Radiomethyl Iodide

The reaction of methyl-C^{14} iodide with a secondary amine has frequently been used to prepare labeled tertiary amines:

$$C^*H_3I + R_2NH \longrightarrow R_2NC^*H_3 \cdot HI$$

In this procedure the reactants, in a suitable solvent, such as ethanol, are allowed to stand for relatively long periods of time. Often the reaction is run in the presence of a base in order to remove the HI formed. Compounds prepared by this procedure include mepivacaine [13], N-methyldiethanolamine [14], codeine [3], and tropine [15]. N-methyl-C^{14}-2-hydroxyethylpiperidine has also been labeled by this procedure and has been used in turn for the preparation of the phenothiazine tranquilizers, thioridazine [16] and methixen [17]. In their preparation of 1-nicotine-N-methyl-C^{14} by this procedure, Leete and Bell [18] found the yield to be low due to quaternary salt formation. For that matter, quarternization, which is frequently experienced as a side reaction, greatly limits the utility of this method.

The methyl iodide alkylation method presents a reliable route to N-methyl derivatives of aromatic amines. For example, at least two laboratories [19, 20] have prepared labeled dimethylaniline this way. Further, Boissonas, Turner, and du Vigneaud [2] have methylated p-methylaminoazobenzene with radiomethyl iodide, but in this case sodium amide was required as a condensing agent.

The methyl iodide procedure has also proved of value for the preparation of highly hindered amines such as butynamine-N-methyl-C^{14} (21), an amine with interesting hypotensive activity:

$$CH_3-\underset{\underset{\displaystyle CH_3}{|}}{\overset{\overset{\displaystyle CH_3}{|}}{C}}-\underset{\underset{\displaystyle}{|}}{\overset{\overset{\displaystyle C^*H_3}{|}}{N}}-\underset{\underset{\displaystyle CH_3}{|}}{\overset{\overset{\displaystyle CH_3}{|}}{C}}-C \equiv CH \qquad \text{Butynamine}$$

It should also be mentioned that this procedure is satisfactory for the preparation of N-methyl-H^3-labeled amines
from radiohydrogen-labeled methyl iodide.   For example,
(−) cocaine, (+) pseudococaine, and (−) scopolamine have all
been labeled by this procedure [22].

Miscellaneous

There are a few special syntheses which have been reported
that are of interest.   For example, Werner et al. [23] have
synthesized N-methyl-C^{14}-labeled atropine from radiomethyl
amine in a four-step synthesis.   El-Merzabani [24], also starting with radiomethylamine, prepared the interesting alkylating
agent 3,3'-dimesyloxy-N-methyl-C^{14}-dipropylamine in a three-
step synthesis.   The key step in this latter preparation was
the Michael addition of two moles of ethyl acrylate to methylamine to yield dicarbethoxyethyl methylamine.   Mention should
also be made of the preparation of labeled dimethylethanol-
amine by Artom and Crowder [25].   In this reaction N-methyl-
ethanolamine was alkylated with radioformaldehyde, employing
alkaline boric acid as a reducing agent.

METHODS WHICH YIELD EITHER SECONDARY OR
TERTIARY METHYLAMINES

Ethyl Chloroformate Method

A number of years ago Easton et al. [26], described the
following route to N-methylamines:

$$RNH_2 \xrightarrow{\overset{\overset{O}{\|}}{ClCOEt}} RNHCOOEt \xrightarrow{LiAlH_4} RNHCH_3$$

With the development in these laboratories of a satisfactory
method for the preparation of carboxyl-labeled ethyl chloroformate, this has become a promising method for the preparation of N-methyl-labeled secondary and tertiary methylamines.
Carboxyl-labeled ethyl chloroformate was conveniently prepared by reacting equimolar quantities of radiophosgene,
absolute ethanol, and purified s-collidine [distilled from

Ba(OH)$_2$] in toluene at 0°. The preparation of the intermediate carbamate is then effected by reacting a primary or secondary amine with the radioactive chlorocarbonate in toluene at room temperature in the presence of an excess of collidine. The resultant carbamate, which frequently can be isolated as a crystalline solid, is then reduced with lithium aluminum hydride to yield the N-methylamine.

The method has been used to label the antidepressant drug nortriptyline [27]. Attempts to label this secondary amine by several other methods has been unsuccessful. Of course, the method can also be used to prepare compounds labeled with isotopic hydrogen by reducing an unlabeled carbamate with either lithium aluminum hydride-H$_4^2$ or lithium aluminum hydride-H^3. Elison et al. (28) have reported the preparation of morphine-N-methyl-H$_3^2$ by this procedure and have described its interesting pharmacological properties. The double-labeled compound N-methyl-C^{14}-N-methyl-H$_3^2$-3-phenylpropylamine has been prepared in these laboratories (unpublished).

CHCH$_2$CH$_2$NHC*H$_3$
Nortriptyline

$$CH_2CH_2CH_2N \begin{matrix} C*H_3 \\ CH_3^2 \end{matrix}$$

N-Methyl-C^{14}-N-methyl-H$_3^2$3-phenylpropylamine

It seems likely that the ethyl chloroformate method will enjoy wide use in the future for the preparation of isotopically labeled N-methylamines. The observation that chloroformates react directly with some tertiary amines to form a carbamate by elimination of a methyl group [8] suggests that it may be possible to label tertiary amines this way without isolation of the secondary amines.

Reduction of N-methylamides

Labeled methylamines can be prepared by the lithium aluminum hydride reduction of N-methyl-C^{14}- and N,N-di-methyl-C^{14}-amides. The amides in turn are prepared from the carboxylic acid and the labeled amine. The method depends on the availability of appropriate carboxylic acid (often a difficult problem) and on the absence in the molecule of groups sensitive to lithium aluminum hydride. Although not widely used, the method has been of value in these laboratories for the preparation of relatively simple labeled amines such as N,N-dimethyl-C^{14}-2,2-diphenylethylamine.

Reaction of Methylamine-C^{14} or Dimethylamine-C^{14} with an Active Halogen

This method, which is of limited applicability, can be illustrated by the preparation of DL-adrenaline-N-methyl-C^{14} [29]. In this preparation chloroacetopyrocatechol is first reacted with radiomethylamine to form N-methyl-C^{14}-labeled adrenalone. Chemical reduction of the adrenalone-C^{14} then yields racemic-labeled adrenaline, a compound that has proved to be of great value for tracer studies on catechol amine metabolism.

Labeled methylanilines with electron-attracting ring substituents can also be prepared this way. For example, N,N-dimethyl-C^{14}-2,6-dinitro-4-trifluoromethylaniline was readily prepared by reacting radiodimethylamine with 2,6-dinitro-4-trifluoromethylchlorobenzene in ether at room temperature [30].

METHODS FOR THE PREPARATION OF N-METHYL-C^{14} SECONDARY AMINES

Alkylation of Sulfonamides with Radiomethyl Iodide

Tosylates of many primary amines can be alkylated with methyl iodide in the presence of a strong base. Hydrolysis of the resultant methylated tosylate yields an N-methyl secondary amine. N-methyl-2-fluoreneamine [31], dimethylamine [32], and monomethylputrescin [33] have been labeled by this procedure:

$$\phi - SO_2NHCH_3 \xrightarrow[C^*H_3I]{NaOH} \phi SO_2N(C^*H_3)_2 \xrightarrow{HOH} \phi SO_3H + HN(C^*H_3)_2$$

Chemical Demethylation of Labeled Tertiary Amines

The preparation of a secondary amine frequently presents a substantially greater challenge than does the synthesis of the corresponding tertiary amine. Because of the recent increasing interest in secondary amines in pharmacology, A. Pohland and his associates at Lilly Laboratories have been investigating a number of synthetic routes to these compounds. One of the most fruitful approaches has been the chemical demethylation of tertiary methylamines. In addition to yielding secondary amines of interest to the pharmacologist, such a reaction is also of great value to the tracer chemist. It would (1) yield secondary amines needed as starting materials for the synthesis of labeled tertiary amines and (2) serve as a route to labeled secondary amines by the chemical demethylation of the labeled tertiary amine.

One of the most interesting agents for this purpose which has been found by Pohland (personal communication) is diethyl azodicarboxylate. Diethyl azodicarboxylate is a powerful hydrogen abstracting agent (cf. Yoneda et al. [34]). It was found by Pohland and his associates to react with tertiary methylamines to form an adduct which, when treated with acid, yields the demethylated amine. The nature of the adduct is not clear as yet but may have the following structure:

$$RN(CH_3)_2 + \underset{\substack{\| \\ NCOOEt}}{NCOOEt} \longrightarrow \left[\begin{array}{l} RNCH_2-N-COOEt \\ \quad\;\; | \qquad\;\; | \\ \quad CH_3 \quad HN-COOEt \end{array} \right] \xrightarrow{\;Acid\;} RNHCH_3$$

This promising reaction has proved to be of great value. For example, it has been used to prepare the interesting analgesic, noracetylmethadol, both in the unlabeled and labeled forms [10, 35, 36]. Other reagents of use in chemical demethylation have also been described in the literature. Some of these may also be of use in the preparation of labeled methylamines.

OTHER POSSIBLE METHODS FOR THE PREPARATION OF LABELED METHYLAMINES

There are, of course, in the organic chemical literature many other routes to methylamines. A few of these which might have application to labeling are mentioned below. For example, Cope et al. [37] have described the preparation of tertiary amines in good yields by the reduction of methiodides of dimethyl tertiary amines with lithium aluminum hydride. Thus, it should be possible to form the methiodide of

a tertiary amine with radiomethyl iodide and then regenerate
the tertiary amine in labeled form by reduction:

$$RN(CH_3)_2 + C^*H_3I \longrightarrow RN (C^*H_3)_3I^- \xrightarrow{LAH} RN(CH^*_3)_2 + C^*H_4$$

Of course, some radiocarbon would be lost as methane, but
the overall process results in the preparation of labeled tertiary
amine without the necessity of preparing the corresponding
secondary amine, often a difficult process.

Pichat et al. [38] have described the following synthesis of
secondary amines:

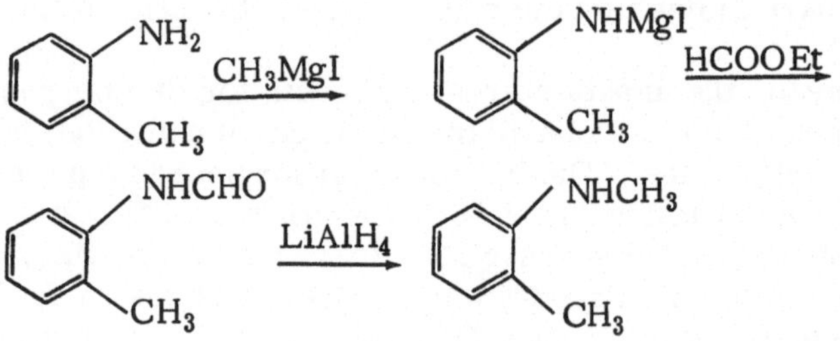

Preliminary work in our laboratory indicates that good yields
based on ethyl formate-carboxyl-C^{14} are obtained in this reaction. Another possible route to labeled secondary amines is
that of Buck [39] who prepared N-methylhomoveratryl amine in
79% yield by the alkylation of the corresponding Shiff's base
with methyl iodide.

Sekiya and Ito [40] have very recently described a new
synthesis of monomethylanilines which appears to represent a
very attractive labeling route, utilizing radioformaldehyde as
a starting material. Their two-step process is as follows:

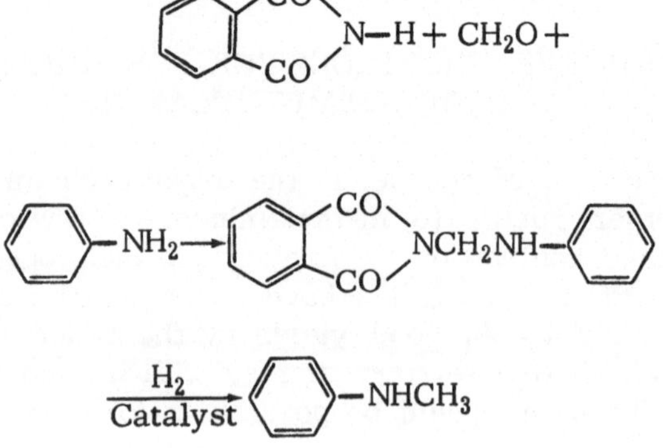

Finally, brief mention should be made of biosynthetic methods. In recent years a variety of methyl transferases have been described, mainly from the laboratory of Axelrod [41]. These enzymes transfer the methyl group from S-adenosylmethionine to an appropriate acceptor. Since methyl-labeled S-adenosylmethionine is commercially available, it seems likely that a number of interesting amines can now be radiomethylated enzymatically. Indeed, the enzyme phenethanolamine N-methyl transferase [42, 43] has been used by Fuller in Lilly Laboratories (personal communication) for the preparation of 1-adrenaline. The obvious advantage of this method over the chemical method described earlier is that the enzymatic procedure produces the natural enantiomorph while the chemical procedure yields racemic product.

SUMMARY AND COMMENT

It is clear from this review that there are now available a variety of methods for the preparation of N-methylamines. The great therapeutic importance of this class of compounds insures that further research will produce still more advances in methodology.

Radiocarbon-labeled methylamines are of great value in pharmacological research. By far the most important pathway of metabolism of these compounds is by oxidative N-demethylation [44]. In the intact animal the end product of the N-demethylation of N-methyl-C^{14} drugs is radiocarbon dioxide. With the current availability of excellent radiorespirometers (cf. Tolbert and Cozzetto [45] and Okita [46]) the rate of metabolism of these drugs can be readily determined by monitoring the rate of appearance of expired radiocarbon dioxide.

Problems which can be approached in this matter include—rate of metabolism, rate of absorption, effect of inhibitors, effect of inducers, effect of nutrition, sex difference, effect of different pharmaceutical formulation, and so forth. In the case of certain drugs, such as morphine [47] and methohexital [48] in which in vivo N-demethylation is unimportant, the N-methyl group represents a stable label for general metabolism and distribution studies. In in vitro experimentation, of course, the N-methyl-C^{14} label is a perfectly stable label as long as the liver microsomal oxidase system is absent.

One difficulty which arises during the use of N-methyl-labeled drugs should also be mentioned. This difficulty arises from the fact that a small amount of the radiocarbon which arises from the oxidation of the N-methyl group becomes involved in the one-carbon metabolism of the body and appears to enter the nucleic acid pool. This fact means that N-methyl compounds in general are unsuitable for lengthy studies in patients. The problem of long-lived residues of this type have been discussed by a number of authors [10, 20, 49, 50].

REFERENCES

1. Tarpey, W., Hauptmann, H., Tolbert, B.M., and Rapoport, H., J. Am. Chem. Soc. 72:5126 (1950).
2. Boissonas, R.A., Turner, R.A., and du Vigneaud, V., J. Biol. Chem. 180:1053 (1949).
3. Rapoport, H., Lovell, C.H., and Tolbert, B.M., J. Am. Chem. Soc. 73:5900 (1951).
4. Hansson, E., and Schmiterlöw, C.G., J. Pharmacol. 137:91 (1962).
5. Ross, J.L., Young, R.L., and Maass, A.R., Science 128:1279 (1958).
6. Ott, D.G., in: Organic Synthesis with Isotopes, Interscience, New York, 1958.
7. Andersen, K.S., and Woods, L.A., J. Org. Chem. 24:274 (1959).
8. Flynn, E.H., Murphy, H.W., and McMahon, R.E., J. Am. Chem. Soc. 77:3104 (1955).
9. Pohland, A., Sullivan, H.R., and McMahon, R.E., J. Am. Chem. Soc. 79:1442 (1957).
10. McMahon, R.E., Culp, H.W., and Marshall, F.J., J. Pharmacol. 149:436 (1965).
11. McMahon, R.E., J. Med. Pharm. Chem. 4:67 (1961).
12. Woods, L.A., Mellett, L.B., and Andersen, K.S., J. Pharmacol. 124:1 (1958).
13. Kristerson, L., Hoffman, P., and Hansson, E., Acta Pharmacol. Toxicol. 22:205 (1965).
14. Foreman, W.W., Murray III, A., and Ronzio, A.R., J. Org. Chem. 15:119 (1950).
15. Fodor, E., Janzso, G., Otvos, L., and Benfi, D., Ber. 93:268 (1960).
16. Zehnder, K., Kalberer, F., Kreis, W., and Rutschmann, J., Biochem. Pharmacol. 11:535 (1962).
17. Aebi, H., Sauber, E., Lehner, H., and Michaelis, W., Arzreimittel Forsch. 14:92 (1964).
18. Leete, E., and Bell, V.M., J. Am. Chem. Soc. 81:4358 (1959).
19. Miller, E.C., Plescia, A.M., Miller, J.A., and Heidelberger, C., J. Biol. Chem. 196:863 (1962).
20. Berenbom, M., Can. Res. 22:1343 (1962).
21. McMahon, R.E., and Easton, N.R., J. Pharmacol. 135:128 (1962).
22. Werner, G., and Mohammad, N., Ann. 964:157 (1966).
23. Werner, G., Schmidt, H.L., and Kossner, E., Ann. 644:109 (1961).
24. El-Merzabani, M.M., Chem. Pharm. Bull. (Japan) 13:1362 (1965).
25. Artom, C., and Crowder, M., J. Am. Chem. Soc. 74:2412 (1952).
26. Easton, N.R., Lukach, C.A., Nelson, S.J., and Fish, V.B., J. Am. Chem. Soc. 80:2519 (1958).
27. McMahon, R.E., Marshall, F.J., Culp, H.W., and Miller, W.M., Biochem. Pharmacol. 12:1207 (1963).
28 Elison, C., Elliott, H.W., Look, M., and Rapoport, H., J. Med. Chem. 6:237 (1963).
29. Cf. Chem. Engr. News, February 8, 1960, p. 51.
30. Marshall, F.J., McMahon, R.E., and Jones, R.G., J. Agr. Food Chem. 14:498 (1966).

31. Little, J.N., and Ray, F.E., J. Am. Chem. Soc. 74:4955 (1952).
32. Searle, N.E., and Cupery, H.E., J. Org. Chem. 19:1622 (1954).
33. Marer, W., Neumann, D., Schröter, H.B., and Schutte, H.R., Z. Chem. 6:341 (1966).
34. Yoneda, F., Suzuki, K., and Nitta, Y., J. Am. Chem. Soc. 88:2328 (1966).
35. Pohland, A., Sullivan, H.R., and Lee, H.M., Absts. 136th Mtg. Am. Chem. Soc. p. 15-0, September 1959.
36. Fornefeld, E.J., Sullivan, H.R., and Thompson, W.E., U.S. Patent 3,213,128.
37. Cope, A.C., Cigonek, E., Fleckenstein, L.N., and Meisinger, M.A.P., J. Am. Chem. Soc. 82:4651 (1960).
38. Pichat, L., Audinot, M., and Monnet, J., Bull. Soc. Chim. 85 (1954).
39. Buck, J.S., J. Am. Chem. Soc. 52:419 (1930).
40. Sekiya, M., and Ito, K., Chem. Pharm. Bull. (Japan) 14:1007 (1966).
41. Axelrod, J., Pharmacol. Rev. 18:95 (1966).
42. Axelrod, J., J. Biol. Chem. 237:1657 (1962).
43. Fuller, R.W., and Hunt, J.M., Anal. Biochem. 16:349 (1966).
44. McMahon, R.E., J. Pharm. Sci. 55:457 (1966).
45. Tolbert, B.M., and Cozzetto, F.J., in: Preparation and Bio-Medical Application of Labeled Molecules, Euratom, 1964.
46. Okita, G.T., in: Isotopes in Experimental Pharmacology, University of Chicago Press, Chicago, 1965.
47. Way, E.L., and Adler, T.K., The Biological Disposition of Morphine and Its Surrogates, World Health Organization, Geneva, 1960.
48. Welles, J.S., McMahon, R.E., and Doran, W.J., J. Pharmacol. 139:166 (1963).
49. Ober, R.E., J. Labeled Compds. 2:203 (1966).
50. Rosenblum, C., in: Isotopes in Experimental Pharmacology, University of Chicago Press, Chicago, 1965.

THE DESIGN OF CRITERIA OF PURITY FOR LABELED COMPOUNDS

Pincus Peyser

Biochemistry Department
New England Nuclear Corporation
Boston, Massachusetts

I. INTRODUCTION

The ultimate requirement of purity for radiochemicals depends upon the design of the particular experiments in which they are employed. The maximum level of impurities, regardless of their nature, that can be tolerated must be significantly less than the minimum change to be detected. Should they be greater, the labeled compound may still be validly employed, if it can be demonstrated by suitable control experiments that the impurities do not contribute to the observed change.

However, since experimental design varies so widely, it would appear to be impractical and unrealistic to expect that levels of purity can be routinely achieved by commercial suppliers that would meet every investigator's requirements. Despite this, the extensive demand for radiochemicals would attest to the success achieved in satisfying most investigators' needs. Nonetheless, the need persists for at least minimum standards of purity for radiochemicals that would receive wide acceptance, as in the case of ACS specifications for reagent grade chemicals [1], and that are now in the process of being established by the National Academy of Sciences for biochemicals [2]. Until such time that a recognized, independent body is organized to establish such standards for radiochemicals, the standards that are currently in use are those that

In addition to being presented at the Eleventh Symposium on Advances in Tracer Methodology, this paper was also presented at the Second International Symposium on Methods of Preparing and Storing Marked Molecules, Brussels, Belgium, November 28–December 2, 1966.

have been largely arrived at independently by the various radiochemical suppliers as a result of their own experiences.

The supplier considers the question of purity in the context
of what impurities are likely to be present initially in the
starting material(s), or are likely to arise during the course
of synthesis. This serves as a guide in selecting procedures
for the isolation and purification of the product, and at the same
time it provides criteria for ascertaining its purity. The investigator, on the other hand, is guided by his particular experimental conditions in designing suitable controls to ensure that
the observed incorporation of radioactivity into an end product
is not due to (1) an impurity serving as a precursor, or (2) a
contaminating artifact accompanying the end product during its
isolation and preparation for isotopic assay, sometimes
referred to in laboratory parlance as "high background."

For the purpose of the present discussion, the design of
criteria of purity of radiochemicals will be considered in the
context of the method of preparation rather than the ultimate
use of radiochemicals in tracer experiments. The subject will
be treated in two parts. The first part will discuss the design
of criteria of purity based on the preparative procedure
employed, and the second part will consider the formation of

Table I.   Factors Which Contribute to the Presence
of Impurities

---

1. The purity of the precursor(s) employed; for example, synthesis of
   DL-leucine-1-C^{14}.

2. The nature of the precursor(s) employed; for example, enzymatic
   synthesis of thymidine-2-C^{14}.

3. The side reactions which may occur (e.g., because of the presence
   of a contaminating enzymes); for example, enzymatic synthesis
   of uridine diphosphate glucose [glucose-C^{14}, uniformly labeled
   (u.l.)].

4. The simultaneous formation of a variety of products; for example,
   biosynthesis of cytidine-2-C^{14}.

5. The methods of isolation and purification employed; for example,
   L-amino acids-C^{14}, u.l.

6. Possible breakdown products due to chemical instability; for example, nucleoside triphosphates-C^{14}, u.l.

---

impurities during handling and storage and its bearing on the design of criteria of purity.

In selecting criteria of purity, consideration is first given to the factors inherent to the preparation of the labeled compound, which may contribute to the presence of an impurity. Some of these are listed in Table I with appropriate examples which will be discussed below in more detail. Having thus arrived at a list of what impurities might likely occur, the techniques subsequently employed in establishing purity are designed to detect their presence and quantitate their level.

II. FACTORS TO BE CONSIDERED IN DESIGNING CRITERIA OF PURITY

A. The Purity of the Precursor(s) Employed—e.g., Synthesis of DL-Leucine-1-C^{14}

This compound may be synthesized according to the following reactions:

$$(CH_3)_2CHCH_2CHO + NH_4C^{14}N \rightarrow (CH_3)_2CHCH_2CH_2(NH_2)C^{14}N + H_2O$$

isovaleraldehyde   DL-a-amino isocapronitrile

$$(CH_3)_2CHCH_2CH_2(NH_2)C^{14}N + 2\ H_2O \xrightarrow{H^+} (CH_3)_2CHCH_2CH_2(NH_2)C^{14}OOH + NH_3$$

DL-leucine

The presence of any DL-a-methylbutyraldehyde in the isovaleraldehyde would give rise to both, DL-isoleucine-1-C^{14} and DL-$allo$-isoleucine-1-C^{14}. Since the boiling points of both these isomeric aldehydes are very close [4], this seemed to be a likely possibility.

The product obtained was subjected to chromatography on strongly acidic ion-exchange resin-loaded paper [3], and 30 to 40% of the radioactivity was found to have a mobility corresponding to that of isoleucine and $allo$-isoleucine. The product was subsequently purified by ion-exchange column chromatography on Dowex 50, H^+, using an increasing concentration of HCl, 0.25 to 1.25 N, for elution [5]. Chromatography of the purified product indicated that it contained less than 0.20% isoleucine or $allo$-isoleucine.

B. The Nature of the Precursor(s) Employed—e.g., Enzymatic Synthesis of Thymidine-2-C^{14}

This compound is prepared enzymatically by incubating thymine-2-C^{14} with an excess of deoxyguanosine in the presence of a dialyzed crude extract of *E. coli* w_c- containing the appropriate nucleoside phosphorylase(s) [6]. The sequence of reactions is as follows:

$$\text{Deoxyguanosine} + \text{inorganic phosphate (trace)} \rightleftharpoons$$
$$\text{guanine} + \text{deoxyribose-1-phosphate} \tag{a}$$

$$\text{Thymine-2-}C^{14} + \text{deoxyribose-1-phosphate} \rightleftharpoons$$
$$\text{thymidine-2-}C^{14} + \text{inorganic phosphate} \tag{b}$$

$$\text{Thymine-2-}C^{14} + \text{deoxyguanosine} \rightleftharpoons$$
$$\text{thymidine-2-}C^{14} + \text{guanine} \tag{a + b}$$

The thymidine-2-C^{14} is isolated by partition column chromatography [7]. The likely radioactive contaminants are thymine-2-C^{14} and possibly thymine-2-C^{14} riboside. The latter could arise if the deoxyguanosine were contaminated with guanosine. The amount of these contaminants is determined by chromatography of a known amount of thymidine-2-C^{14} in a suitable solvent system. The resulting chromatogram is put through a Packard Radiochromatogram Scanner, Model No. 7201, and the areas under the curve corresponding to the positions of thymine and thymine riboside measured and compared with areas of standard amounts of thymidine-2-C^{14} chromatographed in the same manner. In a typical preparation, the product contained less than 0.20% thymine-2-C^{14} and less than 0.10% thymine-2-C^{14} riboside.

The likely nonradioactive contaminants are guanine and deoxyguanosine. Their presence would be indicated by a lowering of the specific activity of the thymidine-2-C^{14}, based on its absorbtivity at 260 mμ, as compared to the starting thymine-2-C^{14}, and by any significant deviation from the ratios of the absorbance at 250, 280, and 290 mμ, respectively, to the absorbance at 260 mμ, which are accepted as characteristic of pure thymidine [2].

C. The Side Reactions Which May Occur—e.g., Enzymatic
Synthesis of Uridine Diphosphate Glucose
(Glucose-C^{14}, u.l.), (UDPG)

Glucose-C^{14} (u.l.)-1-phosphate $+$ UTP $\xrightarrow{\text{UDPG pyrophosphorylase}}$

UDPG (glucose-C^{14}, u.l.) $+$ inorganic pyrophosphate

The product first obtained by the above reaction was pure
on the basis of paper chromatography, ultraviolet spectra, and
the agreement of its specific activity with that of the starting
material. However, upon consideration of the biological
reactivity of UDPG, and that the pyrophosphorylase preparation
was only partially purified, it was suspected that the above
criteria of purity might be inadequate. The product, therefore,
was assayed enzymatically using UDPG dehydrogenase which
catalyzes the following reaction:

UDPG $+$ 2 NAD^+ $\longrightarrow$ UDP glucuronic acid $+$ 2 NADH

The reaction in the presence of excess NAD^+ goes to completion and can be quantitated by measuring the amount of NADH
formed spectrophotometrically at 340 mμ [8]. It was found that
the product gave only 70% of the expected yield of NADH, sug-
gesting that a contaminating nucleotide sugar was present.

Hydrolysis of the product and chromatography of the hydrolysate revealed that the contaminating nucleotide sugar was
uridine diphosphate galactose. This was subsequently demonstrated to have arisen as a result of the presence of uridine
diphosphate galactose-4'-epimerase in the enzyme preparation.
The procedure for the preparation of UDPG pyrophosphorylase
[9] since has been modified so as to minimize contamination by
epimerase. The paper chromatographic systems described by
Dankert et al. [10], for the separation of ADPG and ADP
galactose, will also separate UDPG and UDP galactose. This
system is now employed as a criteria of purity of the product.

D. The Simultaneous Formation of a Variety of Products—e.g.,
Biosynthesis of Cytidine-2-C^{14}

Cytidine-2-C^{14} is prepared microbiologically by incorporation of uracil-2-C^{14} into E. coli w_{c-} [6], a mutant requiring uracil
for growth. The radioactivity is incorporated primarily into

the pyrimidines of RNA and DNA. Repeated extraction of the
harvested cells with 70% (vol/vol) ethanol at 47°C yields a
residue of nucleic acids and protein, which is incubated with
1 N KOH for 18 hr at 37°C to hydrolyze the RNA present to the
2' and 3' ribonucleotides. These are separated from the
undegraded DNA and protein by the addition of three volumes of
alcohol, at pH 2 in the cold, whereupon the DNA and protein
precipitate. The ribonucleotides obtained in the supernatant
are degraded to their corresponding nucleosides employing
E. coli phosphatase. The pyrimidine nucleosides are separated
from each other and the unlabeled purine nucleosides by
partition column chromatography [7]. In designing the criteria
of purity of the isolated cytidine-2-C^{14}, the method of preparation and isolation was taken into consideration. The likely
radioactive contaminants are cytosine-2-C^{14} and the corresponding 2' and 3' ribonucleotides, and uracil-2-C^{14} and its
corresponding ribonucleoside and nucleotides. Although DNA
should not be degraded by alkali, the paper chromatographic
systems selected could also detect any contaminating deoxy-
cytidine-2-C^{14}. The presence of significant amounts of any
unlabeled purines or their derivatives would be revealed by
ultraviolet spectrophotometry.

E. The Methods of Isolation and Purification Employed—e.g., L-Amino Acids-C^{14}, u.l.

These are isolated from an alga, *Chlorella pyrenoidosa*, ATCC
No. 7516, that has been grown photosynthetically in an atmosphere of carbon-C^{14} dioxide as the sole carbon source [11].
After more than 98% of the carbon-C^{14} dioxide has been consumed, the cells are harvested, washed, and extracted in the
following manner:

1. 75% ethanol, 47°C, 30 min, three times
2. 75% ethanol:ether (1:1), 47°C, 15 min, two times
3. Absolute alcohol, ether, and dried
4. 10% (wt/vol) NaCl, 100°C, 8 hr
5. 5% (wt/vol) trichloroacetic acid, 90 to 95°C, 30 min
6. Alcohol, ether, and dried

The final residue is mainly denatured protein. This is
hydrolyzed with 5.5 N HCl at 110°C for 24 hr. As high as 75 to
80% of the total C^{14} in the hydrolysate can be accounted for in

the fifteen pure amino acids isolated. Allowing for losses, and breakdown during hydrolysis, it is estimated that as much as 90% of the residue before hydrolysis may be protein.

Of the eighteen commonly occurring amino acids, three—tryptophan, methionine, and cysteine—are present in *Chlorella* protein in relatively small amounts [12] and usually do not survive acid hydrolysis. Tryptophan appears to be destroyed completely, methionine is oxidized to methionine sulfoxide, and cysteine is oxidized to cysteic acid; occasionally small amounts of cystine and *meso*-cystine are isolated.

The isolation of the amino acids in the hydrolysate is accomplished by ion-exchange column chromatography by a modification of procedures described in the literature [5, 13] as outlined in Fig. 1.

The methionine sulfoxides are eluted from the last column and appear before the threonine peak. No methionine has ever been detected, but if any were present it would be expected to appear between valine and isoleucine, as would cystine and *meso*-cystine which do sometimes occur. The cysteic acid which accounts for most of the cysteine of the protein is separated on the first column since it is not adsorbed to Dowex 50, H⁺. Tryptophan has never been detected in the acid hydrolysate, but if it were present it would be expected to be eluted after histidine.

This brief description of the preparation and isolation of C^{14} uniformly labeled amino acids is presented to provide the

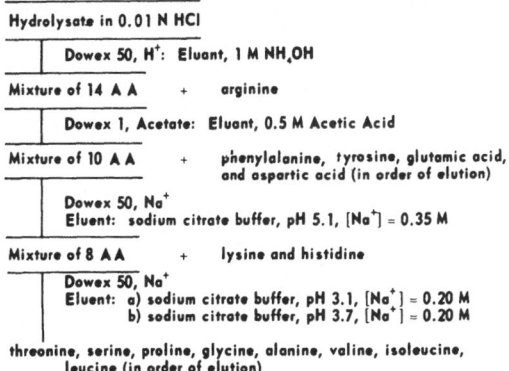

Fig. 1. Scheme for the separation of amino acids-C^{14}, u.l.

necessary background for a discussion of the design of criteria of purity for these compounds. The first and foremost consideration in selecting paper chromatographic systems for the determination of purity is their capability of detecting cross contamination between an amino acid and those that are eluted immediately before and after and between any other contaminants that are likely to occur.

For example, isoleucine is eluted from the column after valine and before leucine, and although it is well separated from these two amino acids, the possibility of cross contamination nonetheless exists. Moreover, as mentioned previously any methionine and cystine present in the protein hydrolysate would be expected to be eluted from the column between valine and isoleucine. It has been reported [13] that during hydrolysis epimerization of L-isoleucine may occur, giving rise to D-*allo*-isoleucine which would be eluted from the column shortly before the isoleucine. Therefore, the systems employed in ascertaining the purity of the isoleucine are designed to separate these possible contaminants, namely, valine, leucine, methionine, cystine, and *allo*-isoleucine. For separating the three leucine isomers, a system employing a strong acid ion-exchange resin-loaded paper [3] was devised. In the six years we have been engaged in the preparation of C^{14}, u.l., L-amino acids, no instance of cross contamination between isoleucine and leucine has been encountered, nor has the presence of an *allo*-isoleucine contaminant been detected.

F.   Possible Breakdown Products Due to Chemical
Instability—e.g., Nucleoside Triphosphates-C^{14}, u.l.

Although consideration of purity of products which are chemically unstable might be more appropriately considered under the topic of stability, there is justification in briefly considering the questions at this point for the following reasons:

1.   The chemical instability of a product often plagues the chemist during its initial isolation and purification.

2.   Although in some cases it is difficult to prevent breakdown from occurring during the various laboratory manipulations, conditions of storage may sometimes be found that stabilize the product.

The example chosen is a rather obvious one, that is, the break-down of nucleoside triphosphates to their corresponding diphosphates and to a much lesser extent to their corresponding monophosphates. Therefore, our paper chromatographic methods of evaluating radiochemical purity are selected on the basis of their capability to separate the nucleoside diphosphates and monophosphates from the corresponding triphosphates.

The examples cited illustrate the rationale employed in our laboratories for selecting criteria of purity of radioactive chemicals which is based primarily on the method of preparation. These may not always be adequate for the investigator either qualitatively or quantitatively. Qualitatively, an impurity could be present that had gone undetected and the possibility of its presence unsuspected. Quantitatively, the level of purity required for certain investigations may be unduly difficult and costly to achieve, and if achieved it is often difficult to maintain for any considerable period of time.

Table II. Steroids: Rates of Decomposition [17]—Stored at 5°C in Benzene—ethanol 9:1 (vol/vol)

Compound	Specific activity (c/mmole)	Time stored (month)	Rate (%/month)
Cortisone-1, 2-H^3			
preparation 1	30	2	5
preparation 1	30	3	0.8
preparation 1 repurified	2	8	0
Estradiol-6, 7-H^3			
preparation 1	30	1	5
preparation 2	35	5	6
preparation 3	42	4	0
preparation 4	5.6	2	0
Estrone-6, 7-H^3			
preparation 1	50	6	10
preparation 2	30	1.5	2.7
preparation 2 repurified	30	2	0
preparation 3	42	3	0.8
preparation 4	2.8	5	0.2

III. DECOMPOSITION OF RADIOCHEMICALS

Thus far in this discussion little has been said about the maintenance of purity. This perhaps is the most troublesome aspect of the entire problem of radiochemical purity. In order to rationally design criteria of purity for compounds that may undergo decomposition during storage or as a result of handling, it is necessary to identify the products that may arise. In this respect our knowledge is rather limited, and the systems used for ascertaining any changes in purity with the passage of time are primarily those employed for the initial determination. If by chance observation they are found to be inadequate, more suitable systems are substituted wherever possible.

Most of our studies, until recently, on the decomposition of radiochemicals have been focused upon the determination of rates of decomposition from the practical view of ascertaining the shelf life of various labeled compounds. At the same time, the results of these studies have provided us with some interesting insights into the scope of this problem, and some suggestions for further investigations. Table II presents some data on the rates of decomposition of some tritium-labeled steroids. (This data was kindly provided by Dr. Geller, head of our Steroid Laboratory.)   Of particular interest is the variation in the rates of decomposition of different preparations of estradiol-6,7-H^3 (No. 1-3) and estrone-6,7-H^3(No. 1-3). Repurification, in some instances, of some rather unstable preparations (cortisone-1,2-H^3, preparation No. 1, and estrone-6,7-H^3, preparation No. 2) resulted in very stable products. These variations could not be attributed to differences in conditions of storage or in specific activity.   Such inconsistencies have been observed with other classes of compounds, which has led us to suspect that the differences in stability of different preparations of the same compound may be due to differences in the nature and amounts of trace impurities present. These might arise from solvents or from the supports used in the various types of chromatography employed for isolation and purification. Indeed, Dr. Geller found that some of the nonradioactive but colored fractions collected during the column purification of some steroids, when added to the pure product, markedly enhanced the rate of breakdown. These preliminary results are extremely suggestive, but as yet insufficient as a basis for any definite conclusions.

The rates of decomposition of twelve C^{14}, u.l., α-amino acids

Table III. α-Amino Acids-C^{14}, u.l.: Rates of Decomposition—Stored at 5°C at a Concentration of 100 μc/ml in Dilute HCl

Amino acid	Specific activity (mc/mmole)	Solvent	Time stored (month)	Rate (%/month)
L-alanine	123	0.01 N	11	<0.01
L-arginine	223	0.01 N	4	1
	240	0.01 N	6	0.44
L-aspartic acid	164	0.01 N	18	0.16
	148	0.01 N	4.5	0.20
L-glutamic acid	200	0.01 N	10.5	<0.04
	205	0.01 N	13	0.11
	209	0.01 N	2	0.33
glycine	80	0.01 N	6	0.43
L-isoleucine	222	0.01 N	4	1
	240	0.01 N	6.5	0.70
	240	0.01 N	6	0.63
L-leucine	275	0.01 N	5	1
L-lysine	247	0.01 N	4	0.30
L-phenylalanine	393	1.0 N	1.5	0.20
	334	1.0 N	1	1.80
L-serine	120	0.01 N	5.5	0.42
L-threonine	156	0.01 N	3	0.56
	160	0.01 N	16.5	0.13
L-tyrosine	351	1.0 N	4	0.35
	334	1.0 N	8.5	0.93
	334	1.0 N	1.5	1.20

are listed in Table III. Six of these were found to decompose at rates of less than 0.5% per month. Where different preparations of a given amino acid have been studied, considerable variation in the rates of decomposition were observed, as in the case of arginine, glutamic acid, isoleucine, phenylalanine, threonine, and tyrosine, with one exception, aspartic acid. As in the examples of the steroids previously cited, these variations cannot be attributed to differences in conditions of storage or to differences in specific activity.

Table IV. Sulfur-Containing Amino Acids-C^{14}, u.l.—Rates of Decomposition

Amino acid	Specific activity (mc/mmole)	Solvent	Concentration (μc/ml)	Temperature (°C)	Time stored (month)	Rate (%/month)	Product identified
L-cystine	190	0.1 N HCl	100	5°	2.5	1.2	cysteic acid
	190	0.1 N HCl	100	-15°	2.5	1.2	cysteic acid
	190	0.1 N HCl	50	5°	4	1.5	cysteic acid
	190	0.1 N HCl plus 0.1% L-methionine	50	5°	4	0.7	cysteic acid
L-methionine	224	0.1 N HCl	100	5°	3	2.9	methionine sulfoxide
	224	0.1 N HCl plus 0.2% DTT*	100	5°	3	1.1	methionine sulfoxide
	198	0.1 N HCl	100	-15°	3	3.1	methionine sulfoxide
	198	0.1 N HCl	100	5°	2	5.2	methionine sulfoxide
	198	0.1 N HCl plus 0.2% DTT	100	5°	2	1.2	methionine sulfoxide

*DTT = dithiothreitol.

Table V. L-Tyrosine—Rate of Decomposition

Amino acid	Specific activity (mc/ mmole)	Solvent	Concentration (μc/ml)	Temperature (°C)	Time stored (month)	Rate (%/month)
L-tyr-C^{14}, u.l.	333	1.0 N HCl	200	−15°	3.5	1.18
	334	1.0 N HCl	100	5°	8.5	0.93
	334	1.0 N HCl	100	5°	1.5	1.20
	351	1.0 N HCl	100	5°	4	0.35
	351	0.50 N HCl	100	5°	4	0.15
	351	0.50 N HCl plus 0.2% DTT*	100	5°	4	0.15
	360	0.01 N HCl	100	5°	4.5	3.85
	369	0.01 N HCl	100	5°	4.5	1.90
	393	0.01 N HCl	100	5°	2	2.8
	393	0.01 N HCl plus 0.2% DTT	100	5°	2	2.6
	393	H$_2$O	100	−15°	2	1.9
	393	H$_2$O plus 0.2% DTT	100	−15°	2	1.5
L-tyr-1-C^{14}	26.6	1.0 N HCl	100	5°	1	2.1
L-tyr-3, 5-H^3	5.6 × 10^3	0.10 N HCl	10^3	5°	10	0.68

*DTT = dithiothreitol.

Table IV tabulates the rates of decomposition for the C^{14}, u.l., sulfur amino acids, L-cystine and L-methionine. These are considered separately since they are particularly unstable. In the case of L-methionine, one of the major decomposition products has been identified chromatographically as methionine sulfoxide. The addition of dithiothreitol [14], a potent reducing agent, markedly reduces the formation of methionine sulfoxide as well as other unidentified decomposition products. This compound may be a suitable means of reducing decomposition, where its presence does not affect the outcome of an experiment. It also may be suitable for storing stock solutions for any prolonged period of time, since dithiothreitol can be readily separated from methionine by passage through a strong cation-exchange resin column. Since dithiothreitol readily reduces L-cystine to L-cysteine [14], and because there is the possibility

Table VI. Nucleosides—Rates of Decomposition

Compound	Specific activity (mc/mmole)	Solvent	Concentration (μc/ml)	Temperature (°C)	Time stored (month)	Rate (%/month)	Product identified
5 BrUdR-2-C^{14}	19	H$_2$O	756	5°	0.75	2.3	5 BrU
	19	70% EtOH	100	−15°	15	0.15	UdR
5 BrUdR-6-H^3	5,860	70% EtOH	1000	−15°	7	0.40	UdR
TdR-2-C^{14}	30	H$_2$O (st)*	100	5°	6	0.43	
	30	H$_2$O (st)	100	5°	6.5	0.70	
	30.5	70% EtOH	100	5°	4.5	0.20	
	42.2	70% EtOH	100	−15°	1.5	0.40	
TdR-CH$_3^3$	10,600	H$_2$O (st)	7,900	5°	1.5	5	
	9,200	67% EtOH	3,000	5°	6	0.08	
	14,200	70% EtOH	880	5°	6	0.27	

*st = sterile.

of forming mixed disulfides, this reagent was not considered suitable for stabilizing L-cystine. On the other hand, it was found that L-methionine does provide some protection against the breakdown of L-cystine.

Table V compares the rates of decomposition of L-tyrosine-C^{14}, u.l., L-tyrosine-1-C^{14}, and L-tyrosine-3,5-H^3 of widely differing specific activities and stored under various conditions. As can be seen from the data there is considerable variation in the rates of decomposition of various preparations of L-tyrosine-C^{14}, u.l. This compound appears to be less stable in 0.01 N HCl stored at 5°C than in water stored frozen, or when stored in more concentrated acid. In addition the results do not reveal any correlation between the rates of decomposition and specific activity, position of the label, or the nature of the isotope; however, more data would be required before any definitive conclusion could be drawn.

A comparison of the rates of decomposition of two pyrimidine deoxynucleosides, both C^{14} and tritium labeled, stored under various conditions, is presented in Table VI. The data indicate that 70% ethanol is a much better solvent than water with respect to stability. Of particular interest is the finding that the major radiochemical impurity, arising in ethanolic solutions of 5-bromodeoxyuridine, has an R_f identical with deoxyuridine in three different paper chromatographic systems. In aqueous solution the major radioactive breakdown product was identified chromatographically as 5-bromouracil. The decomposition occurring in aqueous solution is believed to be the result of chemical instability rather than the result of exposure to radiation. This is based on the finding that 20% of the radioactivity of 5-bromodeoxyuridine-2-C^{14}, in an aqueous solution when heated in an autoclave for 20 min, appeared as 5-bromouracil-2-C^{14}. Aqueous thymidine solutions, on the other hand, can be autoclaved without any appreciable breakdown.

In summary of the experimental data in Tables II to VI, the following points emerge. The considerable variation observed in the rates of decomposition between different preparations of the same labeled compound does not permit an accurate prediction of the shelf life of many radiochemicals. On the other hand, the stability of some preparations of compounds of widely different chemical structure suggest that the prevention of decomposition, or at least a considerable reduction in the rates of decomposition, is not an insurmountable problem.

The need for further investigations in this area is quite apparent, and such studies are being pursued more intensively. A better understanding of the causes of decomposition should provide a better basis for its control. More attention should be given to the question of chemical decomposition in an environment free of a source of radiation, since most radiochemicals of high specific activity are stored at concentrations of 10 to 100 μg/ml, and at these dilutions even nonradioactive compounds, usually considered to be stable, may be actually unstable. It is perhaps because radiochemicals provide a convenient and sensitive means of detection of impurities that their decomposition is so apparent. At this point it would be proper to interject a note of caution, since impurities can arise as artifacts of seemingly innocuous laboratory manipulations. For example, glutamic acid can readily give rise to pyrrolidone carboxylic acid under suitable conditions of pH and temperature [15] or may be esterified when slowly evaporated to dryness in the presence of hydroxylic compounds [16]. In conclusion, a knowledge of the stability of radiochemicals and the nature of the impurities that could arise from their breakdown would permit a more rational approach to the design of criteria of their purity.

SUMMARY

A major parameter in the valid application of tracer methodology employing isotopically labeled compounds is the purity of the tracer compound employed. The ultimate requirement of purity rests upon the design of the experiment, which must allow for the nature of the impurities that might be present and the levels that are permissible and yet yield meaningful data from which valid conclusions may be drawn.

As the situation exists today, where many if not most investigators no longer prepare the labeled compounds they employ, but obtain them from commercial laboratories, the criteria of purity that are used by the manufacturer cannot always satisfy each and every investigator's requirements.

Our own experiences with the design of criteria of purity for labeled compounds is discussed. Briefly, the techniques employed are designed to detect and quantitate the impurities that are likely to occur. A "likely impurity" would depend upon (1) the purity and nature of the precursor(s) employed;

(2) the side reactions they may undergo, due to such factors as contaminating enzymes; (3) the simultaneous formation of a variety of products, as in the case of microbial syntheses; (4) the methods of isolation and purification employed; and (5) consideration of possible breakdown products that may arise due to chemical instability. A brief discussion of the problem of stability during handling and storage is also presented.

REFERENCES

1. Reagent Chemicals, ACS, Specifications, 1965, American Chemical Society Committee on Analytical Reagents (in press).
2. Specifications and Criteria for Biochemical Compounds, National Academy of Sciences—National Research Council, Publication 719, 1960.
3. Unpublished data. Chromatography employs Reeve Angel Amberlite Ion Exchange Resin Loaded Paper, SA-2. The paper is washed before use with the developing solvent, sodium citrate buffer, pH 3.7, $[Na^+] = 0.10$ M, and air dried. Chromatography is carried out in a descending fashion for 8 to 15 hr.
4. Beilstein's Handbuch der organischen Chemie, Julius Springer, Berlin.
5. Hirs, C. H. W., Moore, S., and Stein, W. H., J. Am. Chem. Soc. 76:6063 (1954).
6. Nemer, M., J. Biol. Chem. 237:143 (1962).
7. Hall, R. H., J. Biol. Chem. 237:2283 (1962).
8. Strominger, J. L., Maxwell, E. S., and Kalckar, H. M., in Colowick, S. P., and Kaplan, N. O. (editors), Methods in Enzymology, Vol. 6, Academic Press, New York, 1963.
9. Glaser, L., personal communication.
10. Dankert, M., Passeron, S., Recondo, E., and Leloir L. F., Biochem. Biophys. Res. Comm. 14:358 (1964).
11. Erb, W., and Maurer, W., Biochem. Z. 332:388 (1960).
12. Fowden, L., Biochem. J. 50:355 (1951).
13. Moore, S., Spackman, D. H., and Stein, W. H., Anal. Chem. 30:1185 (1958).
14. Cleland, W. W., Biochemistry 3:480 (1964).
15. Das, H. H., and Roy, S. C., Biochim. Biophys. Acta 62:590 (1962); Greenstein, J. P., and Winitz, M., Chemistry of the Amino Acids, Vol. 3, John Wiley, New York, 1961.
16. Ikawa, M., and Snell, E. E., J. Biol. Chem. 236:1955 (1961).
17. L. Geller, unpublished data.

RADIOLYSIS OF BINARY MIXTURES—A STUDY OF THE RADIATION CHEMISTRY OF METHANOL USING LABELED MOLECULES

A. Ekstrom* and J. L. Garnett

Department of Physical Chemistry
The University of New South Wales
Sydney, Australia

The presence of benzene during the radiolysis of methanol leads to typical "protection" curves for the yields of hydrogen and ethylene glycol. Other predominant scavenging products formed during the radiolysis of benzene—methanol solutions include cyclohexadienemethanol, anisole, 1,4-cyclohexadiene, phenylcyclohexadiene, and biphenyl. The yields of these products display a complex benzene concentration dependence, which may be interpreted in terms of various radical reactions involving benzene. This conclusion is confirmed by an examination of the isotope effects associated with the formation of these products in methanol—benzene and methanol—benzene-d_6 solutions.

Constant positive isotope effects are observed with the formation of anisole and biphenyl, while a constant reverse isotope effect is found in the formation of cyclohexadiene-methanol. For 1,4-cyclohexadiene and phenylcyclohexadiene a positive isotope effect is observed in pure benzene, but this isotope effect decreases with increasing methanol concentration in methanol—benzene solutions and becomes reverse below 0.90 mole fraction benzene.

These results may be used to show the significance of radical scavenging processes as distinct from energy transfer in the elucidation of the "protection" mechanism.

*Present address: Dept. of Chemistry, University of Wisconsin, Madison, Wisconsin.

INTRODUCTION

Deuterated compounds have been extensively used in attempts to elucidate the mechanisms of radiation chemistry [1–3], but the isotope effects associated with their use generally have not been investigated in detail. In a previous study [4] it was shown that in the radiolysis of methanol–benzene solutions a number of scavenging products such as anisole, cyclohexadienemethanol, 1,4-cyclohexadiene, and phenylcyclohexadiene are formed. The present manuscript describes the isotope effects associated with the formation of these products when benzene is replaced by perdeuterobenzene.

EXPERIMENTAL

Details of the identification and yield estimation of the scavenging products were described previously [4–5]. Benzene-d_6 was prepared by the catalytic exchange method [6] and analyzed by low-voltage mass spectrometry after purification. The final product was found to contain 98 atom % deuterium.

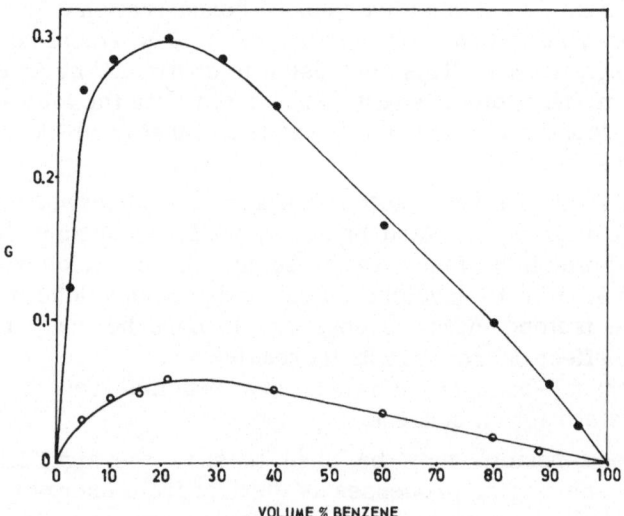

Fig. 1. Effect of benzene concentration on the yields of anisole and cyclohexadienemethanol from methanol–benzene solutions. ● = Cyclohexadienemethanol; o = anisole.

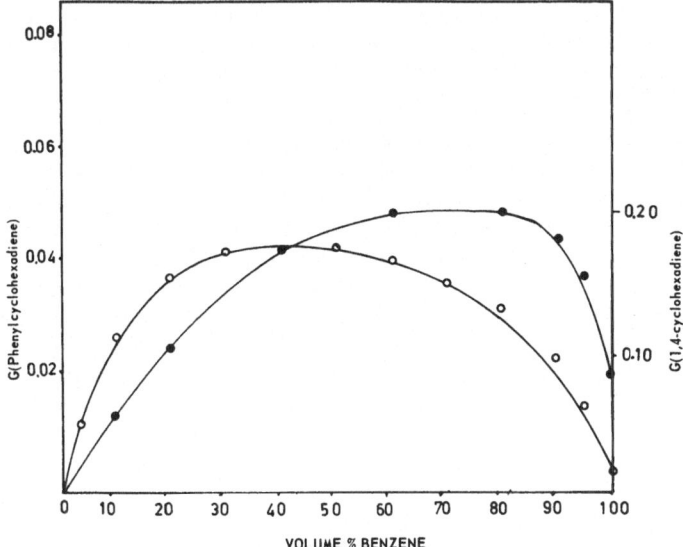

Fig. 2. Effect of benzene concentration on the yields of phenylcyclohexadiene and 1,4-cyclohexadiene from methanol—benzene solutions. • = Phenylcyclohexadiene; o = 1,4-cyclohexadiene.

RESULTS AND DISCUSSION

Figures 1 and 2 show the benzene concentration dependencies of the major scavenging products [4] and Table I summarizes the isotope effects associated with their formation. The results indicate the occurrence of both positive and reverse isotope effects, as well as a transition from positive to reverse isotope effects in the case of 1,4-cyclohexadiene and phenylcyclohexadiene.

The formation of anisole and biphenyl is accompanied by a constant positive isotope effect over the whole range of benzene concentration examined. For both products it is clear that at least one—and in the case of biphenyl, two—C—H or C—D bonds must be broken at some stage, irrespective of the mechanism of formation of these two radiolysis products. In this respect it is of interest to note that the isotope effect associated with the formation of biphenyl is somewhat less than twice the value observed for anisole.

Cyclohexadienemethanol displays a constant reverse isotope effect over the whole range of benzene concentration examined, and these reverse isotope effects must be regarded as "secondary," rather than "primary" isotope effects [7]. It has previously been shown [8] that a reverse secondary isotope effect should accompany a reaction in which the hybridization of a carbon atom under attack is transformed from one of planar sp^2 to one of tetrahedral sp^3. Such a change accompanies the addition of a free radical to a carbon—carbon double bond, and reverse isotope effects have been observed for such reactions [9—11]. Recently, Yang, Scott, and Burr [12] have also observed a reverse isotope effect in the addition of hydrogen and tritium atoms to normal benzene. The reverse isotope effects observed in the present study for cyclohexadiene-methanol are thus consistent with the free-radical addition mechanisms postulated in equation (1).

$$CH_3OH \xrightarrow{\;\gamma\;} H \quad + \quad \cdot CH_2OH \tag{1}$$

The results obtained for phenylcyclohexadiene and 1,4-cyclohexadiene are of interest since a change from positive to reverse isotope effects is observed. Both of the above scavenging products are formed with a small yield in pure benzene, but the yields are sharply increased by the addition of methanol (cf. Fig. 2), and at the same time the isotope effects associated with their formation are sharply decreased.

In pure benzene it is clear that the formation of both

phenylcyclohexadiene and 1,4-cyclohexadiene must involve the addition of hydrogen atoms derived from benzene to benzene molecules. Since this will involve the rupture of a C—H or C—D bond of benzene a positive isotope effect will be expected to dominate [7]. However, upon addition of methanol, the above two scavenging products may also be formed by the addition of hydrogen atoms *derived from methanol* to benzene molecules, and in this case (by analogy with the formation of cyclohexadiene-methanol), a reverse isotope effect is expected. The reactions which may account for the observed results are summarized in equation (2).

From the rapid decrease in the isotope effects associated with cyclohexadiene and phenylcyclohexadiene upon addition of methanol to benzene it may be concluded that

1. The increase in the yield of 1,4-cyclohexadiene and phenylcyclohexadiene observed upon addition of methanol to benzene is due to the scavenging of hydrogen atoms from methanol by benzene.
2. That hydrogen atoms capable of being scavenged by benzene are formed in methanol—benzene solutions containing even high concentrations of benzene.

Table I.   Isotope Effects in the Formation of Radiolysis
                Products in Methanol—Benzene Solutions *

Mole fraction benzene	Biphenyl	Phenyl-cyclo-hexadiene	1,4-cyclo-hexadiene	Cyclo-hexadiene-methanol	Anisole
1.00	2.13	2.25	2.30		
0.90	2.35	1.16	1.10	0.70	1.50
0.81	2.23	1.02	0.95	0.69	1.45
0.65	2.31	0.94	0.90	0.70	(1.23)
0.41	2.23	0.91	0.90	0.75	1.41
0.23	2.67	0.90	0.95	0.70	1.50
0.10	2.21	0.89	0.90	0.82	1.53
0.05	2.19	0.87	0.85	0.80	1.50

$$\text{*Isotope effect} = \frac{\text{Yield of product from methanol—benzene solutions}}{\text{Yield of product from methanol—benzene-}d_6\text{ solutions}}$$

The results thus illustrate the importance of free radical
scavenging mechanisms operating in the present system, and
the use of deuterated additives in the elucidation of radiation
chemistry mechanisms.

ACKNOWLEDGMENT

The authors gratefully acknowledge the support of the
Australian Atomic Energy Commission, which granted study
leave and financial assistance to A. E., the Australian Institute
of Nuclear Science and Engineering which financed the irradia-
tions, and W. Withers, J. Mason, and B. O'Leary for technical
assistance.

REFERENCES

1. Gordon, S., and Burton, M., Disc. Faraday Soc. 12:88 (1952).
2. Burton, M., and Patrick, W.N., J. Phys. Chem. 58:421 (1954).

3. Dyne, P. J., and Jenkinson, W. M., Can. J. Chem. 38:539 (1960); 39:2163 (1961).
4. Ekstrom, A., and Garnett, J. L., J. Phys. Chem. 70:324, (1966).
5. Ekstrom, A., and Garnett, J. L., J. Am. Chem. Soc. 86:5028 (1964).
6. Garnett, J. L., and Sollich, W. A., Australian J. Chem. 14:441 (1961).
7. Halevi, E. A., Progress in Physical Organic Chemistry, Vol. I, Interscience, New York, 1963.
8. Streitweiser, A., Jagow, R. H., Fahey, R. C., and Suzuki, S., J. Am. Chem. Soc. 80:2326 (1958).
9. Matsouka, M., and Swarc, M., J. Am. Chem. Soc. 83:1260 (1961).
10. Takahasi, T., and Cventanovic, R. J., Can. J. Chem. 40:1037 (1962).
11. Mayer, J. R., Copey, W. D., and Robb, J. C., Nature 203:294 (1964).
12. Yang, J. Y., Scott, B., and Burr, J. G., J. Phys. Chem. 68:2014 (1964).

TRITIUM AND CARBON-14 BY OXYGEN FLASK COMBUSTION

Jack D. Davidson and Vincent T. Oliverio

Department of Nuclear Medicine,
Clinical Center and Laboratory of Chemical Pharmacology,
National Cancer Institute,
National Institutes of Health,
Bethesda, Maryland

In the fall of 1961, with the help of Donald Buyske, then of Lederle Laboratories, our group in chemical pharmacology at the National Cancer Institute set up the technique for the oxygen flask combustion of biological samples for the determination of radiocarbon and tritium [1, 2]. After a few months' experience we introduced a number of modifications in the technique from that originally described by Kelly et al. of the Lederle group, and we published the substance of these as a short communication the following summer [3]. During the next 4 years we continued to use this technique and have directly helped about a dozen other groups get started with this method. During these intervening years, a host of publications have appeared, detailing variations on the equipment and technique. Some offer advantages over our technique for certain purposes, but most have impressed us as involving unnecessary complication, expense, or even hazard. The purpose of this presentation is to provide a cookbook version of our entire technique that hopefully will be complete within itself and will make this simplified method of handling biological samples conveniently available to the numbers of people that we think could profitably use it.

The essence of the oxygen flask technique is to provide for the convenient combustion of numerous dried samples of tissues or organic compounds with the quantitative collection of the resultant carbon dioxide and/or water for subsequent liquid scintillation assay of C^{14} or H^3. The practical problem is to suspend the dried sample material in the center of a confined quantity of oxygen and to provide for its ignition and its containment, suspended in the oxygen, with minimal loss of the heat until combustion is complete. It is then necessary to have an arrangement that permits the quantitative collection of the carbon dioxide and/or water into a measured volume of liquid scintillation counting solution.

The very significant contribution from the Lederle group was the standardization upon a 2-liter volume of oxygen and the demonstration that this could combust up to 300 mg of the usual dry organic material samples quantitatively and safely. We have never found it desirable to scale either up or down from this sample size. We have sample bags that contribute only 50 mg of combustible weight to the system and leave 250 mg for the sample itself. In most instances this represents better than a gram of fresh, undried tissue. It might be stressed at this point that this technique is applicable only to samples in which the activity itself is in such a form that it is not volatile and will not be lost incidental to drying the sample preparatory to combustion.

Our technique employs an unmodified 2-liter heavy-walled Erlenmeyer filter flask as the combustion vessel. Such flasks

Fig. 1. Flask for oxygen combustion in Thomas-Ogg igniter apparatus.

are readily available, inexpensive, and tend to survive succes-
sive glass washings better than most of the modified versions
that we have seen (Fig. 1). For a number of years we conformed
to the Lederle group recommendations and used a platinum-
mesh sample holder on a platinum wire stalk, which in turn
was carried by a glass rod through a rubber stopper in the
neck of the flask.   The use of an infrared ignition system
permits this extreme simplification of the "flask head" because
there is no need to provide for insulated, gas-tight electrical
connections to an igniting filament or spark gap within the
flask as are employed in a number of other versions of this
technique.   Our "flask head" is simply a one-hole Neoprene
No. 9 rubber stopper with a suitable length of Pyrex glass rod
mounted in a hole in the stopper and attached to the stalk of
the platinum sample basket.   It is necessary to use Neoprene
stoppers because standard laboratory stoppers are rapidly
deteriorated by the liquid scintillation counting solvents.
Vinyl stoppers are also rapidly attacked by toluene. Silicone
rubber withstands toluene very well but is a bit too slippery to
stay in the flask mouth satisfactorily and is quite expensive.
Closure of the side arm of the flask is simply effected with a
2-in. length of 5.0-mm OD 1.6-mm wall silicone rubber
tubing* and a pair of standard forceps whose jaws have been
covered with short lengths of appropriate gum-rubber tubing to
prevent their sharp edges from cutting the silicone tubing
(Fig. 1).

The design of the sample basket is shown in Fig. 2.   Kelly
et al. described in detail the home fabrication of these platinum
baskets from standard platinum−10% iridium wire and the use
of an electric spot welder [2]. In our short communication we
reported that we could make these baskets with the use of a
Bunsen burner, tack hammer, and a pair of hemostats instead
of the spot welder. The perimeter of the basket was made from
a 12-cm length of 22-gauge platinum−iridium wire shaped into
a square.   The wire ends were held overlapping by clamping
with a pair of forceps, heated to a dull-red glow in the burner
flame, and fused with the tap of a tack hammer against any
convenient hard metal surface.   The remaining wires for the
basket were similarly welded into place using lengths of

*ESCO Rubber Ltd., 34−36 Somerford Grove, London, N. 16, England. Presumably
equivalent tubing as Catalog No. ST650 Meditube is available from Extracorporeal
and Medical Specialties Company, Route 70, Medford Circle, Medford, New Jersey,
08055.

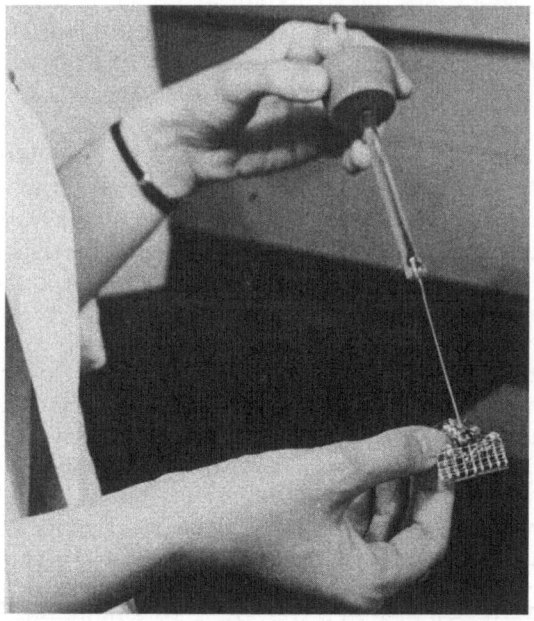

Fig. 2. Combustion bag being inserted in platinum holder of flask head.

22-gauge wire equally spaced, seven in each direction, to form a grid which could pass through the mouth of the 2-liter flask to be employed. This square grid was then mounted on a 15-cm length of 18-gauge platinum—iridium wire to serve as a center support post between the glass rod and the basket. Upon completion, the grid was folded across the center of its horizontal axis to form a trough in which samples could be placed. The wire stalk was finally bent at such an angle as brought the platinum basket within about 2 cm of the wall of the flask with the sample head in position in the flask. The design represents a compromise between minimizing the mass of metal that takes away heat and still having a structure that will prevent the sample from falling to the bottom of the flask and being extinguished.

Commercially made platinum baskets are available at a total cost for platinum and fabrication of about $60 per basket with mounting stalk.* For those with lean budgets, Conway has

*Baker Platinum Division, Engelhard Industries, 113 Astor Street, Newark, New Jersey.

recently advocated the use of nichrome wire baskets [4]. Dr. Donald Reed of Oregon State University reports that he has successfully used nichrome for quite a few years. In our own laboratory we now have had several months' experience with the use of nichrome wire gauze and have encountered no difficulty. There is no question but that nichrome wire will be chemically more reactive than platinum so that it is possible that with certain sample materials, due to their chemistry or higher heat of combustion, some difficulty might be anticipated. Any tritium or carbon that binds to the basket is lost from quantitative recovery. We were able successfully to use the 16-mesh No. 24 B. & S. gauge nichrome wire gauze that is readily available from most scientific supply houses instead of having to go to special suppliers of 8-mesh gauze as reported by Conway [4].

As indicated earlier, we have employed an infrared light beam for ignition of samples. This was chosen over the use of an igniter filament or spark within the flask for two reasons. First, the use of a filament or spark requires a more complicated "flask head" because it must provide for two sealed, lead-in electric wires, and these must be connected to a power source for each ignition. Second, after a certain number of ignitions, which may be of the order of 100, platinum or nichrome filaments burn out. If this occurs during an ignition after the sample has been partially charred, the filament must be replaced and this sample probably has been lost. We found the commercially available Thomas-Ogg infrared igniter* to be extremely effective and convenient. It permitted the use of the simplified "flask head" and gave several thousand ignitions before our first lamp burned out. Furthermore, should a lamp burn out with a sample partially charred, a new lamp can be installed and the ignition completed without unstoppering the combustion flask. The combustion flask sold for use in this igniter is a specialized piece of glassware with a hemi-ball joint on the neck of the flask and with the flask head secured with a mechanical clamp. Apart from being more expensive than a Neoprene stopper, this design, and those using standard taper glass joints, seem to us more hazardous since any excessively fast or large combustion may break the flask instead of popping the stopper as can occur with the Neoprene closure. The Thomas-Ogg igniter provides for aligning the flask and

---

*Infrared igniter, Thomas-Ogg, Catalog No. 6472B, A.H. Thomas Company, Philadelphia, Pennsylvania, 19105.

lamp and has provision for raising or lowering the infrared lamp in order to center the beam on the sample (Fig. 1). It also provides for tilting the flask toward the lamp so that the beam enters the flask at right angles to the flask wall. If you are improvising your own igniter using a Sylvania TruFlector DFC lamp, you will find that it is necessary to have the infrared beam go perpendicularly through the wall of the flask rather than obliquely as would occur with a horizontal beam and the flask standing normally flat on the work surface.

Apart from providing the mechanics and electrical circuit for the ignition, the Thomas-Ogg igniter was designed to provide for containment of any explosive reaction and was accordingly constructed of rather heavy gauge steel walls with a $\frac{1}{4}$-in.-thick acrylic plastic door. We must report most emphatically that these seemingly adequate design considerations have proved insufficient in practice. Due to various

Fig. 3. Igniter apparatus following explosion.

breaches in the procedure, there have been several extremely violent explosions in various laboratories. One such explosion occurred in a laboratory at the National Institutes of Health when ignition using a Thomas-Ogg apparatus was attempted with a flask which apparently contained residual toluene. Figure 3 shows the result. All of us who have had anything to do with this technique of oxygen flask combustion were appalled at the violence of the explosion. The technician who was operating this equipment fortunately sustained only moderate lacerations of her face and neck from the flying glass and partial deafness for a few days. You can see that the plastic door was shattered but that the violence of the explosion was such that even this did not vent it before it had blown out the steel walls and reduced the heavy filter flask to a powder of glass. Thus, this piece of equipment does not afford adequate protection from an explosion. The ignition switch, which in early models was on the top of the cabinet, has been relocated to about 4 ft down the power cord.

In our own laboratory, we now conduct all combustions with the Thomas-Ogg igniter located inside a fume hood with a heavy, nonshatterable glass front window. Our ignition switch has been located on the power cord so that the operator can stand back and to one side at the time of ignition. Before any readers decide that this is no technique for them, may we hasten to reassure you that in our laboratory more than 5,000 combustions have been done with no untoward events, and in another group at NIH the figure is probably 10,000. In the case of the one explosion that did occur at NIH, it seemed possible to reconstruct events to indicate that there was a weak point in the overall procedure. An already used flask containing residual toluene was too readily available to the technician who was responsible for placing fresh samples into clean flasks and flushing them with oxygen. Flask loading and subsequent solvent aliquoting and flask cleaning should be performed in well-separated areas. Ours happens to have been set up in different rooms, and we will deliberately keep it this way.

Samples for oxygen flask combustion are normally prepared in small cellophane envelopes. These can be constructed from short lengths of Visking tubing with the use of Duco or similar "household" cellulose ester cement to glue closed a short fold at the bottom of the length of tubing, thereby forming an envelope. Such Visking tubing envelopes are ideal and, by virtue of the semipermeable nature of the material, afford rapid drying of fluid or wet solid samples. They are, however, laborious to make. A considerable effort has been put into a search for a commercial source of suitable envelopes, and these are now available in the form of rolls of folded cellophane strips with crimped heat seals at 1-in. intervals.* "Bags" can be cut from the roll of 2500 by a simple transverse scissor cut through the middle of the heat seal. The one flap of the bag is slightly longer than the other and permits the use of a mouse ear punch to produce a small hole in the projecting flap so that it can be mounted by a paper clip to an improvised hanger of paper clips arranged on horizontal rods, as shown in Fig. 4. The bag then can be pinched to open it and will accommodate 2 ml of fluid sample, homogenate, minced tissue, powder, snips of chromatographic paper, and so forth. These commercial bags weigh approximately 50 mg and therefore can accommodate up to 250 mg of combustible dried material and still be within permissible limits of the 2-liter oxygen flask system. Drying in the cellophane bags is not as rapid as in the Visking bags because cellophane has a waterproof coating, which is essential to the heat sealing but which does limit water loss to the pathway through the open mouth of the bag. We find, however, that most biological samples of tissue or homogenate will go to dryness overnight or in a couple of days if the bags are hung on a rack in a hood where there is active air flow. Drying can be further accelerated by the use of an array of infrared heating lamps. It is worth emphasizing that the limitation on sample weight to 300 mg of combustible material is due to the gas volume-pressure considerations in this 2000-ml system. The 300-mg total weight can obviously be exceeded to the extent that a proportion of the gross sample weight is contributed by incombustible, inorganic salt.

As has been said before, but cannot be expressed too emphatically, volatile radioactivity is lost incidental to this drying of the samples. It is even difficult to retain C^{14} in the form of carbonate despite appropriate adjustment of fluid samples to an alkaline pH. This is due to exchange of the radioactive carbonate with atmospheric carbon dioxide. Another factor that has worried us is the possibility of bacterial degradation of sample material and loss of activity as a result of its conversion to carbon dioxide. At times when we have been working with samples that contained radioactivity in a compound

*"Combustion Envelopes" from Ivers-Lee Company, 215 Central Avenue, Newark, New Jersey.

Fig. 4. Improvised sample bag drying frame.

that might be very susceptible to bacterial decarboxylation, we have added a few drops of liquid phenol to the samples in bags as a bactericidal agent which is fully combustible itself. With regard to C^{14} in the form of carbonate, we have found that the cellophane bags disintegrate upon the addition of sodium hydroxide but that they will withstand pH 12 borate buffer, which should suffice to retain carbon dioxide if drying can be done in an atmosphere free of carbon dioxide as can be provided by an appropriate desiccator. Fully dried samples, especially heat-dried bags, tend to become brittle, and the bags must be handled carefully to avoid mechanical loss of flakes of sample material incidental to putting the bags into the basket holders for combustion. A wide variety of types of sample material containing both C^{14} and tritium radioactivity have been combusted by ourselves and other laboratories using this technique, with essentially 100% recovery of the radioactivity so long as sample volatity is excluded. We have not encountered the rather consistent quenching that has been reported by some workers [5-7].

There are a few points worth mentioning with regard to the combustion itself. This technique does involve a *combustion* and therefore there must be fuel to be combusted. Some of the failures that we have seen have been directly attributable to the fact that the worker had a sample which had insufficient combustible material to provide the heat required for conversion of the sample to the desired carbon dioxide and water. This means that when the sample size is small, or when it is high in its inorganic salts content, it may be desirable to add fuel. This fuel may be in the form of a sucrose solution to be evaporated with the sample or it may be in the form of some snips of filter paper put in the bag with the sample or even with the sample deposited on the filter paper as a few microliters of sample solution. Samples collected on Millipore filter generally burn exceedingly well because of the cellulose nitrate of the filter. We have found that whole blood samples tend to sparkle at the end of the combustion and have attributed this to their iron content.

Since the infrared ignition depends upon absorption of the light, it is necessary to provide a black felt marker pen dot on the sample bag or to include a small snip of black paper in the bag to assure prompt ignition. We also might belabor the seemingly obvious fact that oxygen is necessary and is provided by putting a hose from a cylinder in the neck of the flask for 10 sec before stoppering. One of the groups that set up our technique had a series of failures due to omitting this step. In the same vein, do not forget to have the clamp on the sidearm tube.

It is desirable to observe each combustion carefully for any signs of technical failure. If there is excessive smoking prior to the true ignition, particulate radioactivity may be deposited in the upper reaches of the flask and subsequently avoid solution in the solvent. Similarly, if any of the sample material falls or drips to the floor of the flask during the combustion, the process is incomplete and quantitative recovery will not be obtained. With this technique as with any other method of combusting organic materials, one does encounter certain compounds which fail to combust quantitatively. This phenomenon is most frequently encountered with compounds having a high percentage of nitrogen or halogen. For this reason it is always good practice to determine on clean samples of your labeled compound that it is amenable to quantitative recovery.

Combustion of a sample is normally complete within less than 1 min. As soon as the flame extinguishes, it is safe to remove the flask from any protective enclosure and set it aside for about 5 min to cool. The next step is the addition of the counting solvent. The composition of this depends upon whether one is dealing with tritium or C^{14}. For the counting of tritium water we employ a solvent mixture of 30% absolute methyl alcohol in reagent grade toluene with 4 gm PPO and 100 mg POPOP/liter. Fifteen ml of this solvent has ample capacity to hold in solution at 0°C the 0.18 ml of water that could arise from the combustion of 300 mg of carbohydrate sample. In practice a volumetric pipette containing 15 ml of the solvent is attached to the silicone rubber sidearm on the flask after the 5-min cooling period, and the clamp is released. Gas pressure in the flask is always negative with respect to atmospheric pressure at this time, and the solvent is drawn into the flask, following which the clamp is replaced on the sidearm tubing. The flask is swirled gently to distribute the solvent over the entire bottom and an inch or two up on the sidewalls. The upper walls of the flask and stopper should not be wetted. The flask is now placed in a cooling bath at −15°C or lower with the cooling confined to the bottom inch of the flask and maintained for 15 to 20 min. We employ a mechanically refrigerated alcohol bath for this purpose, but it can be done equally well using a basin filled with a shallow layer of crushed dry ice in cellosolve. At the end of the condensing period an additional measured 3 ml of counting solvent is delivered again through the sidearm into the flask to rinse in any activity which may have sequestered in the solvent that wetted the sidearm initially. The flask is swirled to mix the solvents. The flask now contains 18 ml of solvent and can be set aside at room temperature for the subsequent removal of a 15-ml aliquot to be placed in a sample counting vial. In the case of C^{14} combustions, including those in which there is both tritium and C^{14}, a scintillation solvent of the composition shown in Table I is employed. Phenethylamine is a good, cheap absorbent for carbon dioxide and provides a higher counting efficiency than Hyamine and better carbon dioxide fixation than ethanolamine [8]. The phenethylamine for this purpose must be redistilled from the material obtained commercially. We do this quite simply by distilling it through a laboratory flash evaporating apparatus with the evaporating flask on a water bath at 95 to 100°C and with an

Table I. CO_2 Scintillation Solvent

Phenethylamine	270 ml
Methanol	270 ml
Toluene	460 ml
PPO	5 g
POPOP	100 mg

aspirator water pump on 12°C or colder water providing a vacuum of about 15 mm mercury.   Such a simple one-plate distillation yields crystal-clear phenethylamine of satisfactory quality for these purposes.   If this is stored in full, brown bottles in the refrigerator, it retains adequate performance characteristics for up to 1 year. The technique for the addition of the solvent for carbon dioxide absorption is exactly the same as that used for tritium water absorption except that in this instance the absorption involves a chemical reaction rather than a condensation and can be permitted to occur at ice-water bath temperature instead of −15°C. The supplemental sidearm rinse with 3 ml more solvent, which can be with or without the phenethylamine ingredient, is again imperative for quantitative recovery.   In the case of either isotope and either solvent system, it is apparent that 15 ml out of 18 or five-sixths of the total combustion products wind up in the counting bottle.

A recent innovation in our laboratory has been to prepare and use solvents without the PPO and POPOP scintillators in them and to provide the scintillators as a final step in the form of 1 ml of a sixteenfold concentrate of PPO and POPOP in toluene added to the 15 ml of solvent in each sample bottle. This keeps the water-insoluble phosphors out of the combustion flasks and considerably facilitates the flask cleaning operation. On the subject of flask cleaning, we find that it can be done very readily using a simple rinse with a warm water solution of common laboratory detergent followed by a tap water rinse. Thorough drying is done in a glass drying oven or by inversion over a metal pipe delivering a jet of air. The air is heated while going through a few coils of the same pipe over a burner prior to delivery to the flask. Under no circumstances should any organic solvent such as acetone be used in a foolhardy effort to expedite drying of the flasks. The danger of a violent explosion from a residuum of such organic solvent in a flask is too great to justify this practice.

The technique presented here uses a minimum of specialized or expensive equipment. The glass components are as rugged as possible. The opportunity for a bigger combustion (euphemism for "explosion") than planned is at a minimum. The technique is not applicable to samples containing volatile radioactivity nor to samples whose net dry weight exceed about 250 mg. It is possibly not the technique of choice if your sample material is limited to 1 to 3 mg, such as can be combusted within the final sample counting vial according to a recent publication [9]. We would like to emphasize that almost any laboratory that is confronted with counting a variety of carbon and tritium labeled samples of biological origin will ultimately realize that combustion is an extremely useful technique. It is the answer to problems with color or insolubility of samples. It is a technique which any laboratory can adopt with a modest expenditure of money and effort.

REFERENCES

1. Kelly, R. G., Peets, E. A., Gordon, S., and Buyske, D. A., Anal. Biochem. 2:267, 1961.
2. Buyske, D. A., Kelly, R., Florini, J., Gordon, S., and Peets, E., Atomlight, No. 20:1, (December 1961).
3. Oliverio, V. T., Denham, C., and Davidson, J. D., Anal. Biochem. 4:188 (1962).
4. Conway, W. D., Grace, A. J., and Rogers, J. E., Anal. Biochem. 14:491 (1966).
5. Dobbs, H. E., Anal. Chem. 35:783 (1963).
6. Conway, W. D., and Grace, A. J., Anal. Biochem. 9:487 (1964).
7. Baggett, B., Presson, T. L., Presson, J. B., and Coffey, J. C., Anal. Biochem. 10:367 (1965).
8. Woeller, F. H., Anal. Biochem. 2:508 (1962).
9. Gupta, G. N., Anal. Chem. 38:1356 (1966).

THIN-LAYER CHROMATOGRAPHY RADIOASSAY: A REVIEW

Fred Snyder

Medical Division*
Oak Ridge Institute of Nuclear Studies
Oak Ridge Associated Universities
Oak Ridge, Tennessee

ABSTRACT

Considerable technological progress has been made in the radioassay of analytical and preparative thin-layer chromatography (TLC) plates since Mangold's original review of this subject (a chapter in Stahl's book, Dunnschicht-Chromatographic, 62–79, 1962). Although the general approach to radioassay of thin-layer chromatograms is similar to that used in paper chromatography, the two techniques require different radiometric procedures because the physical and chemical nature of the materials used in adsorption TLC are completely different from those encountered in partition separations by paper.

Autoradiography, fluorography, strip scanning, zonal scanning, and elution procedures are used for radiometric analysis of thin-layer chromatograms. Thin-layer chromatographic profiles of C^{14} and H^3 determined by liquid scintillation zonal scans are superior to other methods in sensitivity and resolution. Recently the entire zonal-scan procedure was computerized so that the distribution of the chemical mass or radioactivity along a chromatographic lane can be graphed by an electroplotter directed by the TLC computer program. The program gives a direct printout of area percent, recovery, and counting efficiency.

Corroborative TLC (argentation and borate) and gas–liquid chromatography techniques are described for the separation of glyceryl alkyl ethers according to unsaturation, isomeric form, and chain length.

*Under contract with the U.S. Atomic Energy Commission.

INTRODUCTION

The task of quantitatively measuring low-energy β-emitters distributed on thin-layer chromatography (TLC) has been approached in numerous ways (Table I), all of which stem directly from pioneer work in the radioassay of paper chromatography (PC). Mangold's original review [1] of this subject included many of the well-known radiometric techniques then available for PC that might be applied to TLC. However, since these two techniques involve basically different chromatographic processes and use materials differing both physically and chemically, it is not surprising to find that the technology developed in TLC radioassay is strikingly different from that used in PC (Table II). The primary advantage of TLC over all PC techniques is its very large capacity. This permits TLC to serve as a preparative procedure in the purification of specific labeled compounds in milligram quantities, as well as an analytical tool. The loading capacity and open-column features of TLC make it possible to slice narrow zones [2] and obtain a quantitative chemical and radioactive profile of the chromatograms, a technique that is somewhat analogous to column monitoring. The special advantage of TLC zonal analysis is that it measures not only those compounds that would be eluted off columns, but also those compounds that are irreversibly retained on the adsorbent.

RADIOMETRIC SYSTEMS

Elution and Subsequent Counting

The procedure of quantitatively removing an area of the adsorbent from thin-layer plates and subsequently eluting and

Table I. Detection of β-Emitters on Thin-Layer Chromatograms

1. Elution and subsequent counting	4. Fluorography (scintillation autography)
2. Strip scanning	
A. Thin-window Geiger-Müller tube	5. Sublimation and distillation autography
B. Gas-flow Geiger-Müller tube	6. Zonal scanning (liquid scintillation)
C. Phototube-scintillation	
3. Autoradiography	7. Combustion

Table II. Primary Differences in Radioassay of Thin-Layer Chromatograms Compared to Paper Chromatograms

	TLC	PC
Primary process	Adsorption (except reverse-phase, ion-exchange, acrylamide, and Sephadex layers)	Liquid—liquid partition (except adsorbent impregnation)
Capacity	$\mu g \rightarrow mg$	μg
Radioassay		
1. Strip scans	2π	4π
2. Autoradiography	essentially the same	
3. Liquid scintillation	Scrapings	Cut segments

counting the compound isolated has been limited largely to preparative TLC. The basic procedure consists of scraping the zone of adsorbent containing the resolved component into a mound and collecting it on a fritted disc (pore size 20 to 25 μ) in a glass aspirator attached to a water vacuum. After the adsorbent is collected, the aspirator is inverted so that a suitable eluting solvent can be poured over the adsorbent. A number of glass aspirators based on the one described by Goldrick and Hirsch [3] are now commercially available* (Fig. 1).

Another procedure used in elution analysis is simply to scrape with a razor blade a band or zone of the adsorbent containing the isolated compound into a sintered glass or Millipore filter funnel attached to a vacuum flask, so that large volumes of solvent can be passed very rapidly over the adsorbent particles. Although recoveries are generally based on the compound isolated in the eluting solvent, when this procedure is used, a convenient check can be made by distributing the residual adsorbent from the filter on a glass plate, spraying with H_2SO_4, and charring on a hot plate to determine the amount of elemental carbon retained.

The choice of solvents for any elution must be based on the eluotropic series of solvents, the polarity of the resolved compound, and the nature of the adsorbent. Increment radioassay of the chromatogram by elution analysis requires a wide

*Available from Kopp Laboratory Supplies, Inc., 70-13 35th Road, Jackson Heights 72, New York, or Kontes Glass Company, Vineland, New Jersey.

Fig. 1. Adsorbent collection tube (1 to 2 ml capacity) with sintered glass disc (a) for subsequent elution of compound isolated by thin-layer chromatography. Reproduced by permission of Kontes Glass Company, Vineland, New Jersey.

range of solvents, each having different polarity. For this reason, elution analysis is seldom used to determine the quantitative distribution of polar materials along an entire chromatographic lane. When elution analysis is required, as for preparative TLC, it is important to establish whether the chosen eluting solvents will completely recover the resolved compounds from the adsorbent. This is determined by chromatography of a representative small quantity of the compound in those particular solvents. If the compound has an R_f of 0.8 to 1.0 in the system, one can predict that the solvents in question can quantitatively elute it from the adsorbent. However, if the compound remains at the origin or at an R_f lower than 0.8, the solvent is a poor choice; one should choose a more polar eluting solvent and perhaps use additional measures such as steam deactivation of the adsorbent before solvent extraction.

Since compounds prepared by elution analysis are generally used for further chemical or biological studies, it is extremely important to establish that molecular alterations have not occurred during chromatography, scraping, aspiration, elution, and removal of solvents, and that the final product is not contaminated with impurities arising from the adsorbent, solvents, glassware, and filters. Ultrapure adsorbents and solvents are now commercially available, but prior purification of the materials used for most preparative work may be necessary.

Purity of eluted compounds is established by rechromatographing a portion in the chromatographic system used for resolution. If impurities are found, their source must be established. Alterations occurring during the chromatographic procedure itself can be detected by running the original compound (mixture) first in one direction and then in a second direction at right angles to the first in the same solvent system; if no alteration has occurred during development, all components will be on the diagonal formed by a connecting line between the origin and the junction of the two solvent fronts. If alteration during the chromatography is ruled out as a factor in breakdown, one must evaluate the basic steps in elution analysis as a source of molecular alteration. In our experience with elution analyses, the crucial time has been the period during which the compound on the adsorbent is dry and exposed to the atmosphere, that is, while detecting the zone, scraping the zone, and transferring the adsorbent. If breakdown occurs during this time, special care must be taken to keep the adsorbent layer wet with solvent or water vapor and in an inert gas atmosphere. It is also wise to confirm molecular structure; TLC (adsorption chromatography) can detect only altered molecules or contaminants belonging to classes of compounds different from the one represented by the material being prepared.

Strip Scanning

The detectors used in strip scanners for radioassay of thin-layer chromatograms are thin-window G-M tubes, gas-flow G-M tubes, and phototubes in conjunction with NaI crystals [4] or organic scintillators. Many of the commercial scanners originally intended for paper radiochromatograms have been modified [5-11] to accept glass plates of various dimensions. Some workers [12] circumvented these modifications by using adsorbent layers on very thin plastic (now available com-

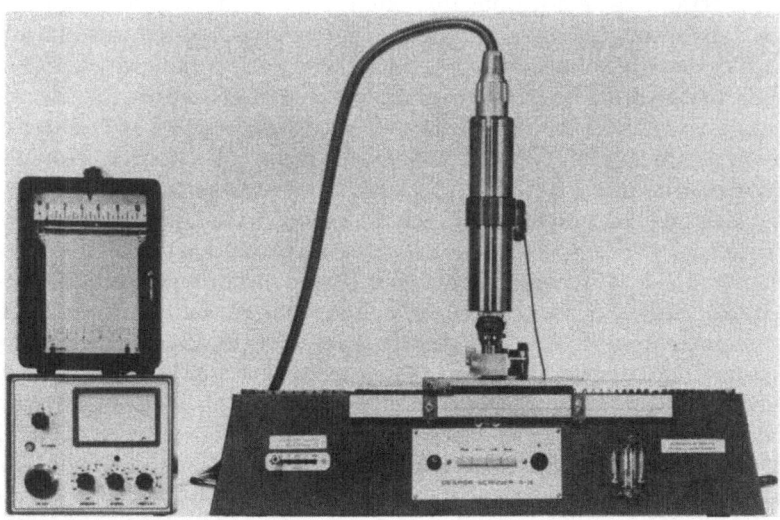

Fig. 2. A two-dimensional radiochromatogram scanner for thin-layer chromatograms. The instrument is manufactured by Laboratorium Prof. Berthold, Schwarzwald, Federal Republic of Germany, and is available in the United States from Brinkmann Instruments, Westbury, New York. Reproduced by permission of Brinkmann Instruments.

mercially),* or by removing the adsorbent on Scotch tape from glass plates after spraying with plastics† that harden the layer [13]. Regardless of the technique used in scanning thin-layer chromatograms by G–M detectors, one is faced with the fact that only 2π geometry is possible. Additional losses of efficiency result from absorption of energy within the adsorbent layer. A commercial strip scanner‡ is shown in Fig. 2; it is patterned after the original model specifically designed by Schulze and Wenzel [14, 15] for thin-layer chromatographic radioassay. This model scans two-dimensional chromatograms with a gas-flow G–M tube of extremely flat aperture which allows close contact with the surface of the adsorbent layer. Other companies (Table III) are now marketing strip scanners with essentially the same basic specifications for radiometric

*Available from Distillation Products Industries, Rochester, New York, 14603. Manufactured by Macherey, Nagel Company and available in U.S. from Brinkmann Instruments, Inc., Westbury, New York, 11590.

†Manufactured by E. Merck (Darmstadt) and available in U.S. from Brinkmann Instruments, Inc.

‡Manufactured by Berthold-Laboratorium and available in U.S. from Brinkmann Instruments, Inc.

measurements.   Both integral (continuous) and increment (stepwise) measurements are possible with these instruments. Efficiencies for C^{14} are quoted as 20 to 30% and for H^3, less than 2%.

Because of the inherent low efficiencies of G–M tube strip scanners, Roucayrol et al. [16, 17] devised an ingenious phototube strip scanner that detects scintillations arising from the interaction between β-emitting compounds and scintillators impregnated in the adsorbent layer before scanning.   Once light quanta are released, problems of self-absorption are eliminated, greatly improving the efficiency for detecting C^{14} (70% efficiency) and H^3 (5% efficiency).   However, the procedure of impregnating the layers of adsorbent with scintillator gel or solution is rather cumbersome and can present problems if the isolated compounds are solubilized during impregnation. Development of a thixotropic scintillator* has recently simplified the impregnation step [18].   The photomultiplier scintillation strip scanner is available from a commercial source† in France (Fig. 3).

In spite of the limitations of strip scanning for quantitative radioassay of labeled compounds possessing low-energy and low-level radioactivity, it can be very helpful in rapidly locating peak zones of activity.   Furthermore, it is a practical method of quantitatively measuring the chromatographic distribution of relatively strong betas or of weak betas of labeled compounds of high specific activity.   Yet neither the resolution nor the sensitivity obtained with strip scans match those obtained by zonal liquid-scintillation analysis.   When our laboratory compared zonal scans and strip scans in their ability to assay a low-level (280 dpm) and a high-level (11,570 dpm) C^{14}-mixture, we obtained identical activity profiles by zonal scan analysis but no peaks at all for the low-level sample assayed by a gas-flow strip scanner operated at peak sensitivity [19].

Autoradiography

Autoradiography of thin-layer chromatograms is an extremely sensitive method for detecting low-energy β-isotopes.

---

*NE221 thixotropic scintillator available from Nuclear Enterprises, 550 Berry Street, Winnipeg 21, Manitoba, Canada.
†Available from SAIP, 38 Rue Gabriel-Crie Malakoff, Paris, France.

Table III. Availability of Commercial Strip Scanners for Thin-Layer Radiochromatograms

Company	Model	Detector	Area size of plate accommodated	Approximate cost
Baird Atomic, Inc. 33 University Rd. Cambridge, Mass.	1-363	GM Gas-Flow windowless	2×8 in.	$3295
Brinkmann Inst.* Cantiague Rd. Westbury, N.Y.	Desaga	GM Gas-Flow windowless	200×400 mm	$3750
Frieseke & Hoepfner GmbH 852 Erlangen-Bruck Fed. Rep. Germany	Not yet available	GM Gas-Flow window-less or GM thin-window (0.9 mg/cm^2)	200×200 mm	Quotation from company
Laboratorium Prof. Berthold 7547 Wildbad Schwarzwald Fed. Germany	LB2027 & LB2725 LB6200	Gas-Flow windowless or with thin-window (0.9 mg/cm^2) Windowless or with thin-window (0.88 mg/cm^2)	up to 200×400 mm 200×200 mm	Quotation from company
Nuclear Chicago 333 E. Howard Ave. Des Plaines, Ill.	1006 (use with model 1002)	GM Gas-Flow window-less or thin-window (0.150 mg/cm^2)	2×8 in.	$3510

Manufacturer	Model	Description	Size	Price
Nuclear Supplies Box 312 Encino, Calif.	EA 1201	GM thin-window (0.07 mg/cm²) and scintillation probe with various crystals	40 × 400 mm	$2750
Packard Inst. Co. 220 Warrenville Rd. Downers Grove, Ill.	7201	GM Gas-Flow windowless or GM thin-window (0.9 mg/cm²)	2 × 8 inch (4 plates can be scanned end to end with special adapter)	$3535
Panax Equip. Ltd. Holmethrope Ind. Estate Redhill, Surrey England	RTLS-1	GM tube with MX113 Mullard ultrathin window or anthracene scintillation detector	20 × 20 cm	$2800
Philips* Eindhoven Netherlands	PW 4004	GM Gas-Flow windowless or GM thin-window (0.9 mg/cm²)	300 × 400 mm	
SAIP 38 Rue Gabriel- Crie Malakoff Paris, France	DMSL 3	Phototube for assay of scintillator-impregnated layers	200 × 50 mm	$4200
Tracerlab 1601 Trapelo Rd. Waltham, Mass.	AD 26 TL attachment to SC525B	GM thin-window GM tube (0.030 mg/cm²)	50 × 200 mm	$3454

*Distributor for manufacturer, Laboratorium, Professor Berthold.

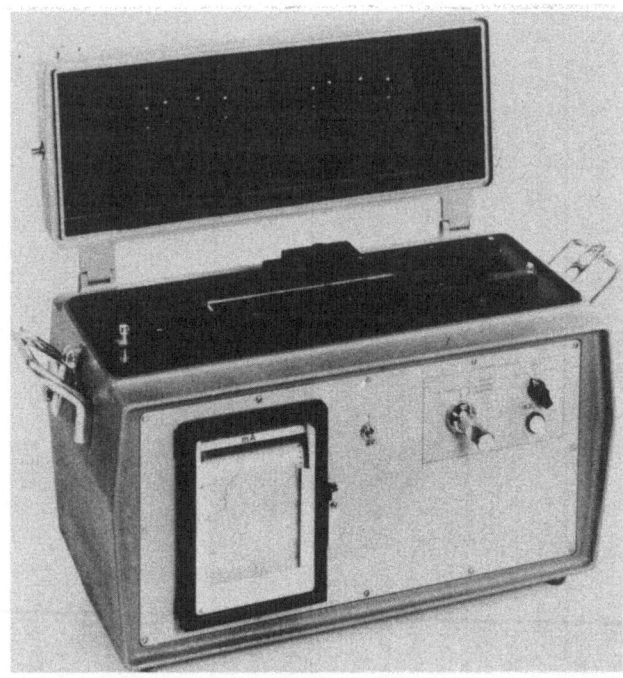

Fig. 3. A scintillation strip scanner (DMSL-3) for thin-layer radiochromatograms impregnated with organic scintillator.   Reproduced by permission of Société d Applications Industrielles de la Physique, Paris, France.

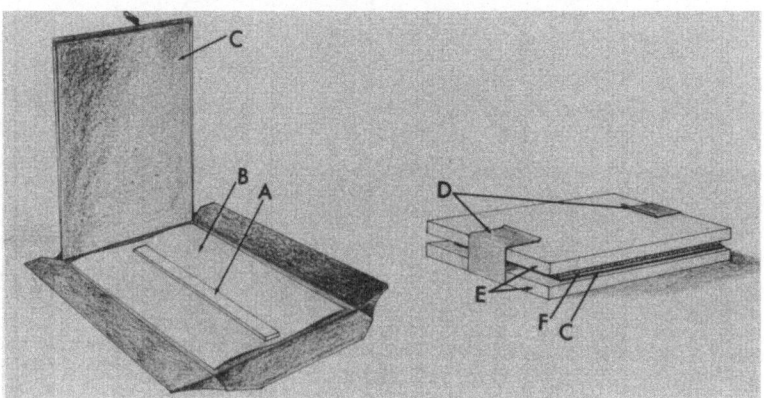

Fig. 4. A holder for X-ray film and thin-layer chromatograms used in the preparation of autoradiograms [23].

However, the relatively poor range of quantitative response for photographic emulsions makes it difficult to measure components possessing different quantities of activity along the chromatographic lane. In addition, the radioactivity is diffused more as the R_f's increase (spots get larger) and self-absorption occurs to a greater extent as the load is increased (the compound is adsorbed deeper in the silica layer). Therefore, this procedure has been limited primarily to qualitative, or at best semiquantitative, radioassay. Guidelines for the approximate time of exposure in the preparation of autoradiograms from TLC plates have been published for C^{14}, H^3, P^{32}, and I^{131} [20].

The technology for autoradiography of thin-layer radiochromatograms is essentially identical to procedures applied to paper radiochromatograms [20-22] with the addition of suitable holders [23-25] (Fig. 4) which provide tight contact between the photographic emulsion and chromatoplates of varying thicknesses. After various periods of exposure of the adsorbent layer to the X-ray film, the film is removed and developed according to standard photographic techniques [23, 26-28]. Since most adsorbent layers are rather fragile, it is generally best to harden the layer by spraying it very lightly with an artist's spray, such as Krylon.* This has little effect on the absorption of even the very low-energy β-emitters such as H^3. When higher-energy β-emitters are being measured by autoradiography, the entire thin-layer plate may be wrapped with a thin plastic sheeting (Mylar†). If chemical interaction occurs between the compounds on the thin-layer (estradiol [24]) and the photographic emulsion, it is mandatory to use a plastic covering. Two isotopes having sufficiently different energies can be distinguished by absorbing the lower energies with a cover sheet of proper thickness.

Autoradiography of H^3-labeled compounds on thin-layer chromatograms' poses special problems because of the short range of H^3 betas (about 1μ) in photographic emulsions [29]. Here it becomes necessary to replace X-ray films with more sensitive emulsions, such as Kodak nuclear track emulsion type NTD [27] and photographic films having high (1250) ASA

*Manufactured by Krylon, Inc., Norristown, Pennsylvania, and available at most artists' supply stores.
†Manufactured by E.I. DuPont de Nemours, Wilmington, Delaware, and available from Brownell, Inc., 85 Tenth Avenue, New York, New York.

ratings [28]. The sensitivity of autoradiography can be increased still further by impregnation of adsorbent layers with nuclear emulsions and by fluorography [29], a procedure described in the next section. Such techniques reduce or circumvent the loss of low-energy betas by adsorption in the protective layer of gelatin that covers the photographic emulsion on commercial sheet film.

Development of specialized techniques is also necessary for preparing autoradiograms of special layers, such as those used in electrophoresis. Lambiotte [30] recently described a thin-layer autoradiographic electrophoretic procedure for measuring H^3-compounds. He made X-ray film suitable for electrophoresis by immersing it in a buffered ionic solution and blotting it between two sheets of filter paper. After electrophoresis he dried the filmstrip and stored it in the dark until development. Since the molecular separations occur in the photographic emulsion, this technique provides the utmost in sensitivity in autoradiography of electrophoretic thin-layer chromatograms.

The darkened areas on a developed autoradiogram can be measured photodensitometrically with the instruments* [23] used to measure carbon residues on adsorbent layers that have been sprayed and charred [31]. A skilled person can measure carbon residues with < 2% error, but good quantitative photodensitometry of autoradiograms is very difficult to attain because of absorption losses within the layer and because of the limited range of film response.

The main advantages of autoradiography for obtaining general chromatographic patterns of radioisotopic distribution are economy, simplicity, and relative speed. In addition, autoradiography provides useful visual confirmatory records and is the method of choice in two-dimensional chromatography [32] and when quenching is encountered in liquid-scintillation radioassay.

Fluorography (Scintillation Autography)

This technique involves the interaction of low-energy β-particles with organic scintillators in the adsorbent layer to produce light quanta traveling into the photographic emulsion and a consequent image on the developed X-ray film. The

*Model available from the Photovolt Corporation, 115 Broadway, New York, New York, 10010.

sensitivity of this technique for H^3-detection is greater than that of regular autoradiography because of decreased self-absorption loss. However, it has been reported that fluorography on paper chromatograms (impregnated with scintillator) is only one tenth as sensitive as impregnation with nuclear emulsions [29].

Fluorographic visualization of paper radiochromatograms was first developed by Wilson [33], who used a liquid scintillation solution of p-diphenylbenzene (terphenyl) to analyze H^3-labeled compounds isolated from Chlorella. A series of more descriptive papers by Wilson [29, 34—36] soon followed his original discovery, but it was not until a report in 1964 by Jolchine [37] that the fluorographic technique was applied to the detection of C^{14}- and H^3-labeled compounds on thin-layer chromatograms. A year later Wilson and Spedding [29] also discussed the general technique of fluorography as a radiometric tool for thin-layer chromatograms. Although supporting data were not given, they claimed that commercial silica gel itself is a scintillator when exposed to ionizing radiation and that its sensitivity is doubled when sprayed with anthracene. Anthracene had been used in paper fluorography [38]; Luthi and Waser [39] extended its application to TLC, incorporating it directly into adsorbent slurries before chromatoplates were coated. Anthracene offered many advantages over other scintillators: low solubility in chromatographic solvents, relative chemical inertness, and very high fluorescent efficiency. These workers [38] also found that the fluorescent intensity of anthracene was considerably enhanced by lowering the temperature, and that this temperature effect was not so great for other scintillators.

Sublimation and Distillation Autography

Wilson and Spedding [29] have discussed the novel technique of subliming and distilling labeled molecules from a chromatogram into a photographic emulsion where they interact to form images after development. In their article, they cite a number of examples supporting their belief that without the investigators' knowledge compounds sublimed during routine autoradiography of paper chromatograms. This could be a significant source of error if comparisons were made between two compounds, only one of which sublimed. Sublimation and

Table IV. Scintillation Solutions for Counting TLC Scrapings

Components	Scintillation Solution I	Scintillation Solution II
A. Phosphor	2,5-diphenyloxazole (PPO) 7.0 g; p-Bis-2-(4-methyl-5-phenylox-azolyl) benzene (POPOP) 0.3 g	2,5-bis-[5'-tert-butylbenzoxazolyl (2')]-thiophene (BBOT) 4.0 g
B. Naphthalene	100 g	80 g
C. Solvents	Dilute components IA & IB to 1 liter with dioxane	Add 600 ml toluene and 400 ml methylcellosolve to components IIA & IIB
Water addition	Mix scintillation solution I (A, B, C,) with water 15:3	Mix scintillation solution II (A, B, C,) with water 30:1

distillation autography is useful only with a limited number of compounds, for example, glycolic acid [29]. There has been no thorough investigation of this method as a practical analytical procedure. However, one can appreciate the problems as well as the virtues that sublimation and distillation might have in conventional autoradiography and fluorography.

Zonal Scanning (Liquid Scintillation)

The scraping of TLC areas into vials for highly efficient suspension [40] or liquid [41, 42] counting by 4π scintillation methods is both quantitative and simple. Multiple isotopic discrimination is also possible within the same sample with liquid scintillation spectrometers. However, when liquid scintillation is used to assay adsorbent particles [40, 42], self-absorption of energy from C^{14} and H^3 adsorbed on the particles or on the surface of the counting vial can be a source of considerable error. The use of plastic vials eliminates the latter problem, but one must establish that the sample does not diffuse into the plastic. A C^{14}-compound that remains on the silica gel particles (10 to 25μ) when homogeneously suspended with a scintillation gel will count approximately 8% less than when it is in solution; and if the adsorbent particles containing an H^3-compound are suspended rather than dissolved, a decrease of 25% occurs in the counting efficiency. The C^{14} self-absorption problem in liquid scintillation solutions can be eliminated by reducing the particle size of the adsorbent to less than 10 μ, well below the range of the C^{14} β-particle [43]. However, the addition of water to scintillation solutions has been the most practical means of eliminating the problem of self-absorption. The water deactivates adsorptive sites and, in effect, solubilizes the labeled compounds; it may also provide penetration of the scintillator into the crystal lattices of the adsorbent particle. Formulations of two deactivating scintillation solutions are listed in Table IV. If the problem of adsorption and self-absorption for certain types of compounds cannot be prevented, formulas [42] must be used to correct for the different counting efficiencies of that portion of radioactivity on the particle and of that portion in solution.

One of the fortunate things about the liquid scintillation radioassay of thin-layer chromatograms is that most of the routine materials used as adsorbents and as detecting reagents do not have any quenching effect in the liquid scintillation system (Table V). Although quenching is encountered after H_2SO_4 charring, visualizing the labeled compounds with fluorescent materials or iodine vapor is almost as effective. Iodine vapor is particularly useful when the iodine crystals are heated on a hot plate and the vapor is allowed to completely saturate the chromatogram; such heavy staining of layers with iodine will visualize most organic compounds. It is not necessary to resublime the iodine from the thin-layer plate; even at high concentrations of iodine, quenching in the liquid scintillation solutions (listed in Table IV) does not occur. Fluorescent visualizing reagents can be either sprayed on the developed chromatogram or incorporated into the adsorbent slurry before preparation of the layer.

Our group soon realized that the scraping of large TLC areas corresponding to a single visualized spot was not ideal for examining the radioactive distribution along a chromatographic lane since chemical and radioactive distributions are not necessarily identical. High specific activity compounds and low-level radioactive peaks can easily remain undetected if chemical visualization techniques serve as criteria for determining the areas to be scraped and counted. For these reasons, our group developed and evaluated a manual [42] and an automatic [2] zonal scraper and collector (Fig. 5) that permits very narrow (1 or 2 mm or larger) continuous zones

Table V. Effect of Adsorbents, Supports,
Complexers, and Indicators on the
Scintillation Process

Nonquenchers	Quenchers
Adsorbents	Alcoholic sprays (e.g., iodine in methanol)
Cellulose	H_2SO_4
Iodine vapor	Elemental carbon
Dichlorofluorescein	Phenylhydrazine
Rhodamine 6G	Ag^+ (12%)
$AgNO_3$, $Ag(NH_3)^{+2}$ (<3%)	
Borate	
Arsenite	
Silicone	
HAc (50λ)	

 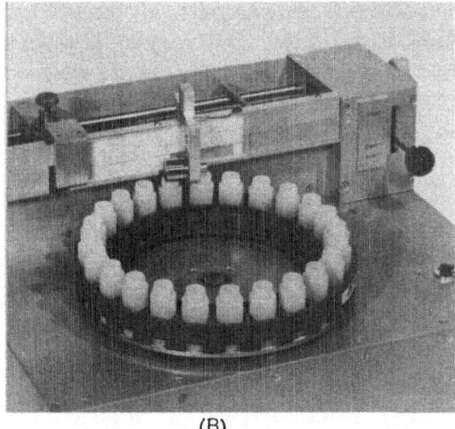

(A)   (B)

Fig. 5. Manual (A) and automatic (B) zonal scrapers for the preparation of chemical and radiometric fractions (1 and 2 mm) for the determination of chemical and radioactive chromatographic profiles (zonal scans). See Snyder and Kimble [2] and Snyder [42] for details of basic components.

along the adsorbent layer to be scraped accurately, quantitatively, and reproducibly. Blueprints* for the original models described in the literature are available and have been used by a number of universities and companies. The original units were designed strictly for analytical zonal analyses of thin-layer plates having a dimension of 2 × 20 cm. These narrow plates could be divided into two lanes, one for the radioactive compound and the other for a nonradioactive standard. Recently, the blade holder has been modified (Fig. 5) so that a standard-size end plate (5 × 20 cm) can be scraped in three or four lanes, one lane at a time. This modification permits the determination of radioactivity on one lane and chemical mass on a second lane; the remaining one or two lanes are used for visualizing the separations by standard H_2SO_4 charring procedures, which can be documented by photography. The slowest procedure in the zonal quantitative radioanalysis of thin-layer chromatograms always has been the tedious calculations and graph-drawing required. Now, however, the entire zonal scan system has been automated [44] so that through data transmission, card punching, computer analysis, and electroplotting we obtain automatically the zonal profile for an entire TLC lane of

*Blueprints available from Clearing House for Federal Scientific and Technical Information, National Bureau of Standards, U.S. Department of Commerce, Springfield, Virginia, 22151.

separated compounds, along with appropriate calculations and statistical data. This system, together with the zonal scraper, has brought the analysis of thin-layer chromatograms up to the level of speed and efficiency inherent in TLC separation.

To be able to include the measurement of mass in the automated procedure would obviously be advantageous, but the photodensitometric [31] and spectrophotometric [45] methods available have not easily lent themselves to such treatment. However, a completely new technique, based on the light-quenching effect of products of an H_2SO_4 reaction [45], has been developed for this purpose (Fig. 6). The details will be published elsewhere but in summary it can be stated that a radioactive source such as Tc^{99} (β^-, 0.29 Mev, $T_{1/2}$= 2.1 × 10^5 years) imbedded in a plastic scintillator is quenched by a sample "charred" in H_2SO_4. This method permits the determination and compilation of chromatographic zonal profiles of chemical mass ranging from five to several hundred micrograms of material with the use of equipment and computer programs already described for radiometric zonal scans.

The resolving capabilities of zonal profile analysis coupled with the high degree of sensitivity of radioactive measurements provide important information on the chromatographic capabili-

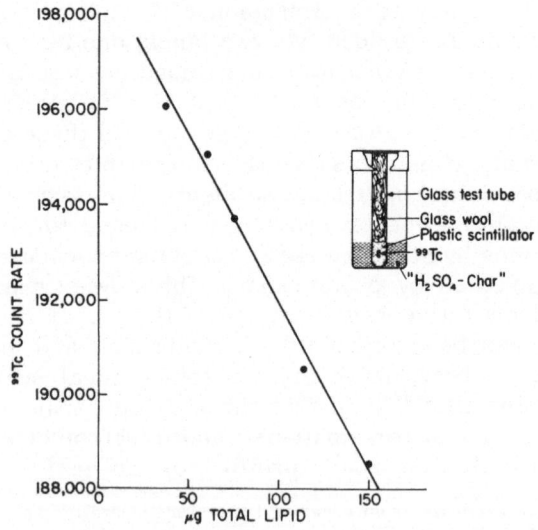

Fig. 6. Liquid scintillation-quenching procedure for quantification of lipid mass on thin-layer chromatograms.

ties of any particular TLC system. In practice, 1- and 2-mm zonal analysis can be used to detect cross contamination of areas caused by tailing from an area of higher R_f [46]. Mathematical methods for evaluating incremental and integral measurements of the distribution of labeled compounds on chromatograms have been applied to TLC zonal scans [47]. They indicate that the zonal scrapers will be very useful in detecting isotopic fractionation [48] in the TLC zones.

Since the development of the zonal scraper, adsorbent layers on plastic film* have become available. The thinner layers of the Eastman product (only $100\,\mu$) limits the loading capacity so that compounds of a low specific activity cannot be detected. A thicker layer ($250\,\mu$) on plastic sheets is now available from Macherey, Nagel. However, the impossibility of cutting identical minute segments and of quantitatively determining chemical mass in the presence of plastic and organic binders makes the use of plastic TLC sheets, at best, a qualitative procedure.

RADIOPURITY

The determination of the radiopurity of commercial compounds perhaps has been the most widely used application of TLC to radioassay; we have reviewed [49] this subject for the radiopurity of lipids and included methods generally applicable to other compounds. Thin-layer chromatography is particularly excellent for establishing radiopurity of nonvolatile compounds in that the contaminants revealed can rapidly be removed by applying the same TLC system on a preparative scale. The presence of impurities must be established under conditions where the compound in question runs at an intermediate R_f, that is, about 0.5. A single chromatographic peak at the origin or at the solvent front is meaningless in evaluating the purity of a compound, yet this is the type of chromatogram that sometimes accompanies the manufactured product. Adsorption chromatography will generally reveal class purity only, and absolute molecular homogeneity can be determined only by complementary techniques (such as gas–liquid chromatography) in addition to TLC.

*Available from Distillation Products Industries, Rochester, New York, 14603. Manufactured by Macherey, Nagel Company and available in U.S. from Brinkmann Instruments, Inc., Westbury, New York, 11590.

SPECIAL TECHNIQUES FOR QUANTIFICATION AND IDENTIFICATION OF MOLECULAR SPECIES

In this section, a limited number of specialized TLC techniques and their application will be discussed. A reference guide to specific types of labeled compounds that have been separated and radioassayed by TLC is given in Table VI.

Organic reactions with labeled reagents are often used so that unlabeled compounds can be quantitatively assayed by

Table VI. Method Papers Dealing with TLC of Labeled Compounds

Type compound	Isotope	References
Amino acids and proteins	C^{14} I^{131}	[30,66] [63]
Carbohydrates	C^{14} H^3	[37,67] [37]
Inorganic ions	Ba^{140}, La^{140}, Ba^{133}, Cs^{133}, Ca^{47}, Sc^{47}, Y^{90}, Sr^{90}, Zn^{72}, Ga^{72}, Nb^{95}, Ta^{182}, Zr^{95}, $S^{35}O_4$, $P^{32}O_4$ Y^{90}, Sr^{90}, Nb^{95}, Zr^{95} Fe^{55} Ca^{45}, Cl^{36} Cl^{36}, Br^{82}, I^{131} Th^{234}, U^{238}, Cl^{38}, S^{35}, P^{32} Na^{24}, K^{22}	[7] [10] [9] [17] [4] [61] [62]
Lipids	C^{14} H^3 P^{32} I^{125} I^{131}	[2,3,5,19,23,26, 28,42,44,46, 47,50] [2,23,26,42,44,50] [2,23] [2] [23,26]
Steroids	C^{14} H^3	[9,24,64,65] [9,14,24,65]
Miscellaneous organic compounds	S^{35} C^{14}	[14] [68]
Nucleic acids and related compounds	C^{14}, P^{32} I^{131}	[32] [11]

radiometric TLC. The two best examples of such reactions are methylation [26, 50, 51] of fatty acids and acetylation [26, 50] of alcohols. These methods are quantitative and offer a high degree of sensitivity for detection. The radioactive products, methyl esters and acetates, can readily be separated from unreacted compounds of other classes by TLC, and then from any homologs in their respective classes by gas—liquid chromatography. The precise quantities of the labeled products can be calculated by isotope dilution methods.

In the technique called separation-reaction-separation chromatography, organic reactions involving synthesis [52] and degradation [53] may even be carried out directly on the TLC layer. This involves the separation of a mixture in one direction, an organic reaction directly on the adsorbent layer, and then the development of the chromatogram in the same system in the second direction. The change in functional groups causes the derivatives to move at R_f's different from those of the original reactants, which, moving at the same rate as they did during the first development, will all lie on a diagonal line between the origin and the junction of the two solvent fronts.

Subfractionation by TLC is used to increase the specificity of GLC analysis, resolving critical pairs not separated by partition chromatography alone. For example, acetoxymercuri-methoxy derivatives [54] and argentation TLC [55] have been used to separate subspecies of a given class according to the degree of unsaturation, and arsenite or borate ions incorporated into the TLC adsorbent layer have been used to separate isomeric forms of dihydroxy compounds [56]. A TLC subfractionation applicable to glyceryl ethers is shown in Fig. 7. First, TLC was used to separate the glyceryl ether diester class from the corresponding triglyceride class (Fig. 7A). Second, argentation TLC [57] was used to separate the deacylated glyceryl ethers on the basis of double bonds (Ag π bond complexes, Fig. 7B). Third, the deacylated glyceryl ethers were separated according to 1- and 2-isomers [44] by TLC on silica gel containing approximately 5% boric acid (Fig. 7C) or by gas—liquid chromatography of the trifluoroacetate derivatives (Fig. 7D) [57]. This GLC procedure can also be used to resolve the glyceryl ethers according to chain length and double bonds (Fig. 7D). Corroborative techniques, such as argentation-TLC [55] and GLC of intact molecules according to carbon number [58], are extremely useful in the structural analysis of triglycerides as well. The position of the fatty acid on the isolated

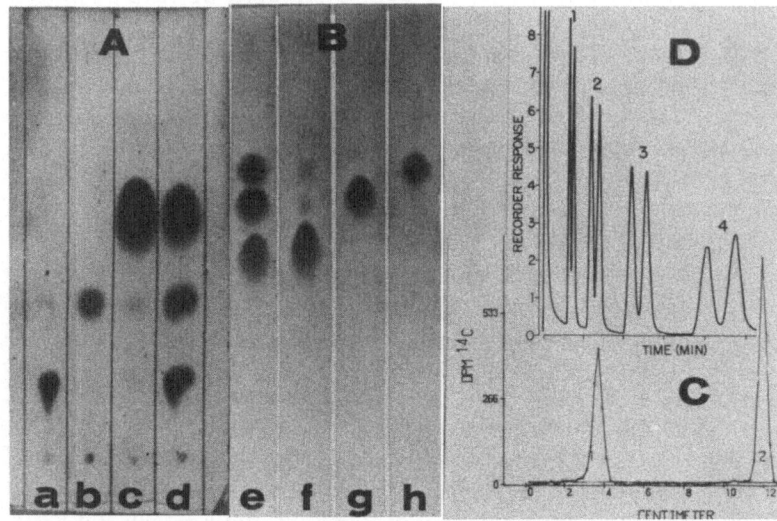

Fig. 7. An application of thin-layer radiometric analysis used in conjunction with gas–liquid chromatography. A. Separation of ether- and ester-containing lipid classes: (a) triglyceride, (b) glyceryl 1-alkyl ether diester, (c) neutral plasmalogen, (d) mixture of a, b, and c, [solvent system, hexane:diethyl ether (95:5 vol/vol)]. B. Separation of a single class of lipids (glyceryl ethers) on argentation layers according to degree of unsaturation: (e) mixture of 18:0-1, 18:1-1, and 18:2-1 glyceryl alkyl ethers, (f) 18:0-1 glyceryl alkyl ether, (g) 18:1-1 glyceryl alkyl ether, (h) 18:2-1 glyceryl alkyl ether, [solvent system, chloroform:ethanol (90:10 vol/vol)]. C. A C^{14} zonal profile of the separation of glyceryl ether isomers on silica gel layers containing boric acid: (1) 1-isomers, (2) 2-isomers, [solvent system, chloroform:methanol (98:2 vol/vol)]. D. A gas chromatogram of the separation of individual glyceryl ethers, as their trifluoroacetate derivatives, according to chain length and isomeric form. The numbered peaks represent the following carbon numbers: (1) 12:0, (2) 14:0, (3) 16:0, and (4) 18:0. The 2-isomers were eluted after the 1-isomer. The data depicted in B and D are from Wood and Snyder [57] and the zonal scan in C from Snyder and Smith [44]. The neutral plasmalogen sample [53] was obtained from H.K. Mangold and H.H.O. Schmid, The Hormel Institute, University of Minnesota, Austin.

triglycerides can be determined with lipase and phospholipase A in conjunction with either organic phosphorylation [59] or diglyceride kinase phosphorylation [60] of the diglycerides formed by lipase.

In conclusion, it is essential to stress that thin-layer chromatography by itself serves limited ends. However, its very effective application to radiometric analysis in inorganic, organic, and biochemistry is fully appreciated when it is used in connection with other established chemical and chromatographic procedures.

ACKNOWLEDGMENT

I wish to thank my wife Cathy for help in preparing this review.

REFERENCES

1. Mangold, H.K., Isotopentechnik, in: Stahl, E. (editor), Dünnschicht-Chromato-graphie, Springer-Verlag, Berlin, 1962.
2. Snyder, F., and Kimble, H., Anal. Biochem. 11:510 (1965).
3. Goldrick, B., and Hirsch, J., J. Lipid Res. 4:482 (1963).
4. Berger, J.A., Meyniel, G., and Petit, J., Compt. Rend. 255:1116 (1962).
5. Rosenberg, J., and Bolgar, M., Anal. Chem. 35:1559 (1963).
6. Moye, C.J., J. Chromatog. 13:56 (1964).
7. Moghissi, A., J. Chromatog. 13:542 (1964).
8. Boucke, G., Atompraxis 11:263 (1965).
9. Bleecken, S., Kaufmann, G., and Kummer, K., J. Chromatog. 19:105 (1965).
10. Breccia, A., and Spalletti, F., Nature 198:756 (1963).
11. Massaglia, A., Rosa, U., and Sosi, S., J. Chromatog. 17:316 (1965).
12. Squibb, R.L., Nature 198:317 (1963).
13. Csallany, A.S., and Draper, H.H., Anal. Biochem. 4:418 (1962).
14. Schulze, P.E., and Wenzel, M., Angew. Chem. Intern. Ed. Eng. 1:580 (1962).
15. Wenzel, M., and Schulze, P.E., Chem. Ing.-Tech. 37:1024 (1965).
16. Roucayrol, J.C., and Taillandier, P., Compt. Rend. 256:4653 (1963).
17. Roucayrol, J.C., Bergner, J.A., Meyniel, G., and Perrin, J., Intern. J. Appl. Radiation Isotopes 15:671 (1964).
18. Roucayrol, J.C., Faculté de Médecine Biologiste des Hôpitaux, Paris, France, personal communication.
19. Snyder, F., Separation Science 1:655 (1966).
20. Fink, R.M., Dent, C.E., and Fink, K., Nature 160:801 (1947).
21. Benson, A.A., Bassham, J.A., Calvin, M., Goodale, T.C., Haas, V.A., and Stepka, W., J. Am. Chem. Soc. 72:1710 (1950).
22. Grennen Tsuk, R., Castro, T., Laufer, L., and Schwarz, D.R., in: Sirchis, J. (editor), Proceedings of the Conference on Methods of Preparing and Storing Marked Molecules, Brussels, November 13–16, 1963, Euratom, Brussels, pp. 497–509, 1964.
23. Snyder, F., Radioisotope Sample Measurement Techniques in Medicine and Biology, IAEA, Vienna, pp. 521–533, 1965.
24. Richardson, G.S., Weliky, I., Batchelder, W., Griffith, M., and Engel, L.L., J. Chromatog. 12:115 (1963).
25. Privett, O.S., Blank, M.L., Codding, D.W., and Nickell, E.C., J. Am. Oil Chemists' Soc. 42:381 (1965).
26. Mangold, H.K., Kammereck, R., and Malins, D.C., in: Cheronis, N.D. (editor), International Symposium on Microchemical Techniques, 1961, Interscience, New York, pp. 697–714, 1962.
27. Sheppard, H., and Tsien, W.H., Anal. Chem. 35:1992 (1963).
28. Fray, G., and Frey, J., Société de Chime Biologique Bulletin 45:1201 (1963).
29. Wilson, A.T., and Spedding, D.J., J. Chromatog. 18:76 (1965).
30. Lambiotte, M., Atomlight 45:10, (May 1965).
31. Privett, O.S., and Blank, M.L., J. Am. Oil Chem. Soc. 39:520 (1962).
32. Simonis, W., and Gimmler, H., J. Chromatog. 19:440 (1965).
33. Wilson, A.T., Nature 182:524 (1958).
34. Wilson, A.T., Proceedings of the Second International Conference on Peaceful Uses of Atomic Energy, Geneva, September 1958, Pergamon Press, New York, pp. 213–214, 1959.

35. Wilson, A.T., Biochim. Biophys. Acta 40:522 (1960).
36. Wilson, A.T., J. New Zealand Inst. Chem. 28:87 (1964).
37. Jolchine, G., Physiol. Vegetale 2:341 (1964).
38. Parups, E.V., Hoffman, I., and Jackson, H.R., Talanta 5:75 (1960).
39. Luthi, U., and Waser, P.G., Nature 205:1190 (1965).
40. Snyder, F., and Stephens, N., Anal. Biochem. 4:128 (1962).
41. Brown, J.L., and Johnston, J.M., J. Lipid Res. 3:480 (1962).
42. Snyder, F., Anal. Biochem. 9:183 (1964).
43. Helf, S., in: Bell, C.G., and Hayes, F.N. (editors), Conference on Liquid Scintil-
    lation Counting, August 20–22, 1957, Pergamon Press, New York, pp. 96–100,
    1958.
44. Snyder, F., and Smith, D., Separation Sci. 1:709 (1966).
45. Marsh, J.B., and Weinstein, D.B., J. Lipid Res. 7:574 (1966).
46. Snyder, F., in: Rothchild, S. (editor), Advances in Tracer Methodology, Vol. 2,
    Plenum Press, New York, pp. 107–113, 1965.
47. Klein, P.D., Separation Sci., 1:511 (1966).
48. Klein, P.D., in: Giddings, J.C., and Keller, R.C. (editors), Advances in Chroma-
    tography, Vol. 3, Marcel Dekker, Inc., New York, pp. 3–65, 1966.
49. Snyder, F., and Piantadosi, C., in: Paoletti, R., and Kritchevsky (editors),
    Advances in Lipid Research, Vol. 4, Academic Press, Inc., New York, pp.
    257–283, 1966.
50. Mangold, H.K., Fette, Seifen, Anstrichmittel 61:877 (1959).
51. Schlenk, H., and Gellerman, J.L., Anal. Chem. 32:1412 (1960).
52. Kaufmann, H.P., and Wessels, H., Fette, Seifen, Anstrichmittel 68:249 (1966).
53. Schmid, H.H.O., and Mangold, H.K., Biochim. Biophys. Acta 125:182 (1966).
54. Mangold, H.K., and Kammereck, R., Chem. Ind. 1032 (1961).
55. Barrett, C.B., Dallas, M.S.J., and Padley, F.B., Chem. Ind. 1050 (1962).
56. Wood, R., and Snyder, F., Lipids, 2:161 (1967).
57. Wood, R., and Snyder, F., Lipids 1:62 (1966).
58. Kuksis, A., and McCarthy, M.J., Can. J. Biochem. Physiol. 40:679 (1962).
59. Brockerhoff, H., J. Lipid Res. 6:10 (1965).
60. Slakey, P.M., Lands, W.E.M., and Pieringer, R.A., Fed. Proc. 25:521 (1966).
61. Seiler, H., and Seiler, M., Helv. Chim. Acta 48:117 (1965).
62. Seiler, H., Helv. Chim. Acta 46:2629 (1963).
63. Massaglia, A., and Rosa, U., J. Labelled Compds. 1:141 (1965).
64. Vahouny, G.V., Borja, C.R., and Weersing, S., Anal. Biochem. 6:555 (1963).
65. Benraad, J., and Kloppenborg, P.W.C., Clin. Chim. Acta 12:565 (1965).
66. Drawert, F., Bachmann, O., and Reuther, K.H., J. Chromatog. 9:376 (1962).
67. Moye, C.J., and Goldsack, R.J., J. Appl. Chem. 16:209 (1966).
68. Kratzl, K., and Puschmann, G., Holzforschung 14:1 (1960).

MODIFICATION AND UTILIZATION OF ISOTOPE FRACTIONATION PHENOMENA IN ANALYTICAL SEPARATIONS

Peter D. Klein, V. Cejka,
and Patricia A. Szczepanik
Division of Biological and Medical Research
Argonne National Laboratory
Argonne, Illinois

A recent review [1] of isotopic fractionation phenomena in analytical separations has shown that there has been a striking increase in the number of reports dealing with these effects in the last 4 years. Whether this increase is due to improvements in analytical techniques or to an improvement in our climate of scientific acceptability cannot be determined, but in either instance their numbers indicate that isotope fractionation is not the uncommon occurrence it was once thought to be. In particular, it has become evident that differences in the chromatographic mobilities of two isotopic species are manifested not only by small molecules but by substances of molecular weight 400 to 600 as well; that evidence of isotopic separations is not limited to processes of high-resolution capability; and that fractionation has been demonstrated in a wide variety of biochemical compounds.

The ultimate explanation of these isotopic fractionation phenomena awaits a fuller theoretical development of chromatographic selectivity, which is in turn influenced by the conditions, types, and magnitudes of the phenomena themselves. Fortunately, the exploration and utility of these isotope fractionations is not limited by the availability of theoretical explanation; they can be used and studied in the same manner as other molecular properties affecting chromatographic mobility. For example, use of this information has been made to improve isotope dilution methodology [2] by correcting for the presence of fractionation in the purification of the derivative;

Work performed under the auspices of the U.S. Atomic Energy Commission.

still other applications have been in the refinement of radiochemical purity and identity criteria [3, 4]. Cejka and coworkers [5] have demonstrated that experimental modification
of fractionation phenomena can be achieved in some partition
systems. By changing the composition of the stationary phase
used to prepare the column, the separation of aldosterone-1,2-
H^3 from aldosterone-4-C^{14} could be varied over a threefold
range and that of cortisone-1,2-H^3 from the -4-C^{14} form over
a tenfold range. These results suggest that a deliberate en-
hancement of the isotopic effect to achieve significant sample
enrichment or its reduction to abolish such selectivity may be
more accessible to the investigator than previously supposed.
Information gained in this manner also serves to prove the
intramolecular forces arising from the presence of the isotopic
atom in a particular region of the molecule. Some examples
of such probes will be illustrated in this report.

Chromatographic retention data may be used to obtain
thermodynamic properties of solute molecules. In certain
cases where the isotopic species are completely resolved by
the column, the change in these properties resulting from
isotopic substitution has been determined. Liberti, Cartoni,
and Bruner [6] have measured the retention values for benzene
and perdeuterobenzene at a variety of temperatures with
squalene and silicone oil as the stationary phases. The relationship between the ratio of the retention volumes (V_r^0) for benzene and perdeuterobenzene and the temperature of chromatography is given by

$$\ln \frac{V_r^0 \, \text{benzene}}{V_r^0 \, \text{d-benzene}} = \frac{(\Delta H_{\text{benzene}} - \Delta H_{\text{d-benzene}})}{R} \cdot \frac{1}{T} + C$$

where ΔH is the enthalpy of the chromatographic process. From
the slope of the temperature dependence, the difference in
enthalpy associated with the chromatographic process between
the two forms of benzene was found to be -46.2 cal/mole
on squalene and -11.4 cal/mole on silicone oil. Using the
difference in the latent heat of vaporization between the two
forms (-20.96 cal/molecule) [7], Liberti and coworkers parti-
tioned the enthalpy differences between the vaporization process
and the process of mixing with the stationary phase. The latter
was thus calculated to be -25.2 and 9.6 cal/mole for the two
phases used.

Saha and Sweeley have studied the free-energy difference
in helium for protium and deuterium sugars as their trimethyl

silyl ethers by obtaining the ratio of the corrected retention times for each sugar [8]. From this ratio, differences of -3.5 and -13.9 cal/mole were found for α-glucose and β-glucose in the protium and deuterium forms, chromatographed on SE 30. Differences related to the stationary phase were also evident since values of -20.5 and -38.3 cal/mole were found for the same pairs of isotopic anomers on Carbowax-20M.

Both of the foregoing studies required at least partial peak disengagement of the two isotopic forms in order to determine the individual retention values. This is because the same peak characteristic (total mass) is used to monitor both components of the emerging peak. Such studies are thus limited by the total number of plates available in the chromatographic system and thereby to particular chromatographic systems, notably gas—liquid, where the total number can be made large. On the other hand, simultaneous measurement of two isotopes in the same peak as provided by dual-label scintillation counting permits a drastic reduction in the number of plates as well as enabling precise determination of separation factors in the range 1.01 to 1.0001 [4]. This extension of the separation factor range makes it possible to determine thermodynamic quantities as small as 0.06 cal/mole.

Figure 1 illustrates the relationship between the retention volume and column temperature for aldosterone-1,2-H^3 and aldosterone-4-C^{14} in a liquid—liquid partition system employing 50% methanol as the stationary phase and toluene/ligroin (4:1) as the mobile phase. From the slope of the line, the heat of solution of aldosterone in 50% methanol was determined to be -3170 ± 95 cal/mole. In a quantity as large as this, the small but consistent difference in partition coefficient displayed by the tritium-labeled aldosterone as opposed to the radiocarbon form cannot be evaluated mathematically because of the relative magnitudes of the total heat of solution and the isotopic increment resulting from the substitution of tritium for radiocarbon in the aldosterone molecule. However, if the C^{14} form is used as the point of reference in each instance, one may obtain the secondary partition coefficient or separation factor as

$$\ln\left[\frac{K_{H^3}}{K_{C^{14}}}\right] = \ln\left[1 + \frac{\Delta M\%}{100}\right]$$

where $\Delta M\%$ is the percent displacement between the two forms computed either from probit analysis or from the isotope ratio

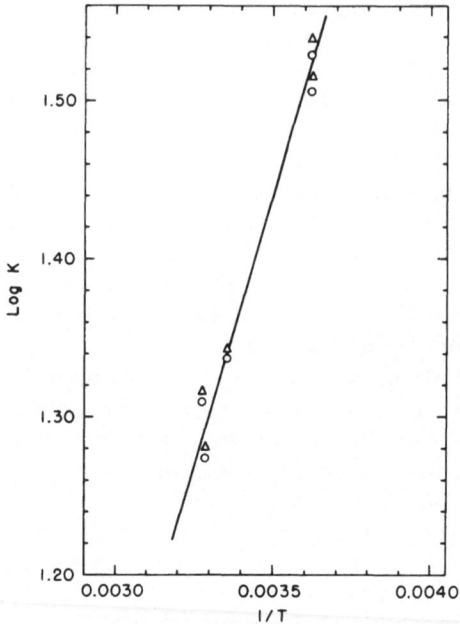

Fig. 1. Standard retention volumes of aldosterone-1,2-H^3(triangles) and aldosterone-
4-C^{14} (circles) as a function of temperature.

[9]. These values are shown in Fig. 2, together with their
standard errors and the slope drawn according to the method
of least squares. This slope gives an incremental heat of
solution due to isotopic substitution of -8.1 ± 2.1 cal/mole,
or 0.025% of the total heat of solution. The sensitivity of the
comparative technique suggests that an assessment of the
intramolecular contributions to the heat of solution is feasible:
Systematic substitution in various regions of the solute molecule and measurement of the effect upon separation factors at
a variety of temperatures is an easily foreseeable utilization
of this principle.

Amino acids are a group of compounds particularly suitable
for such investigation. In fact, one of the earliest reports of
an isotopic effect in a chromatographic separation was that
reported by Piez and Eagle for C^{14} amino acids on ion-exchange
columns [10, 11]. Today, the commercial availability of labeled
amino acids often includes both tritium and radiocarbon
varieties of the same species and, in fact, may include several
alternative labeled sites with the same isotope. It has been our

recent interest to reinvestigate these isotope effects to determine whether tritium substitution produced any significant effect and whether positional factors in C^{14} and H^3 labeling had any effect on the chromatographic mobility of these compounds. To our extreme delight, both effects were evident in a number of amino acids of which several selected examples are shown in Figs. 3–5. In the instance of serine (Fig. 3) it can be seen that there is a strong effect upon the retention volume by the 1-C^{14} substitution and only a slight effect when the radiocarbon is in position 3. On the other hand, introduction of tritium on carbon 3 *increases* the retention volume of the labeled compounds relative to the unlabeled variety. In Fig. 4 the variety of C^{14} glutamic acids now available show three distinct radiocarbon mobility increments, that of -1-C^{14} being the largest, with -5-C^{14} having considerably less effect and that of -3,4-C^{14} being only slightly greater than the unlabeled material. Again there is a strong effect upon the mobility of the amino acid when the 3 carbon is substituted with tritium instead of hydrogen. It has been suggested by Peyser [12] that this may be indicative of a normal interaction with or steric

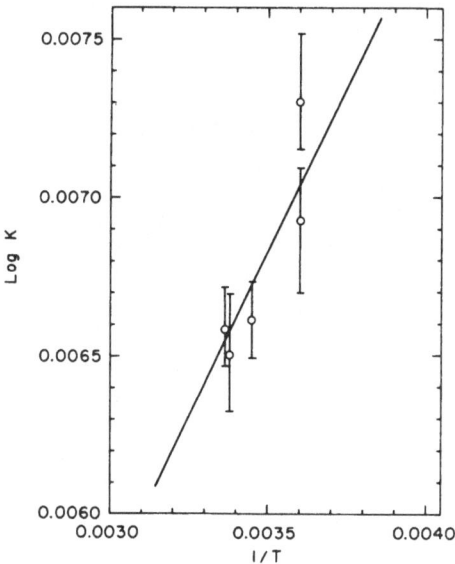

Fig. 2. Separation factors of aldosterone-1,2-H^3 from aldosterone-4-C^{14} as a function of temperature. Bars represent standard error of each measurement.

hindrance of the amino group by this hydrogen position whose
stability is greatly increased by the presence of the heavier
isotope. Other available evidence appears to bear this out, but
molecular models show the interaction distance to be quite
large by any ordinary considerations, and such a mechanism
clearly requires further investigation. Finally, the "chromatographic spectrum" of the five methionine species shown in
Fig. 5 is illustrative of some of the unexpected findings to be
encountered in the study of isotope effects. The substitution of
C^{14} in positions 1 and 2 produces the anticipated retardation in
the retention volume of methionine, whereas labeling the methyl
group with radiocarbon produces a slight shortening or earlier
emergence. In contrast to these three species, the H^3 methyl
methionine shows an inordinately large reduction in the retention volume of more than 0.6%. This is a particularly large
effect for a substituent so far removed from the ionic centers of
the amino acid and invites a number of speculations concerning
its mechanism. One might suppose that during the chromatography of methionine a charge center develops on the sulfur
atom whose magnitude is altered by isotopic substitution in
neighboring atoms. This possibility would imply that the effect
of the C^{14} methyl group, in which the isotopic atom is directly
linked to the sulfur, should be larger than that of the H^3 methyl
group where there is an intervening bond between the isotopic

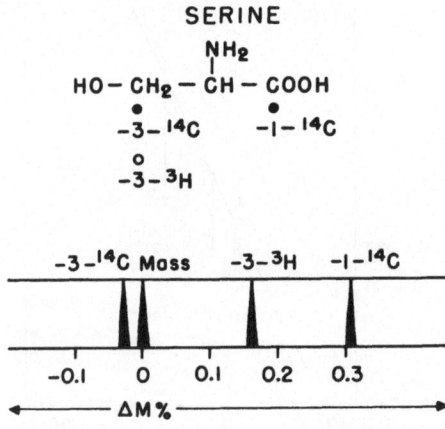

Fig. 3. Isotopic displacement of labeled serine species from mass (ninhydrin).
Center of triangle indicates midpoint of each peak, width is indicative of the
uncertainty of midpoint position, not chromatographic resolution of each form.

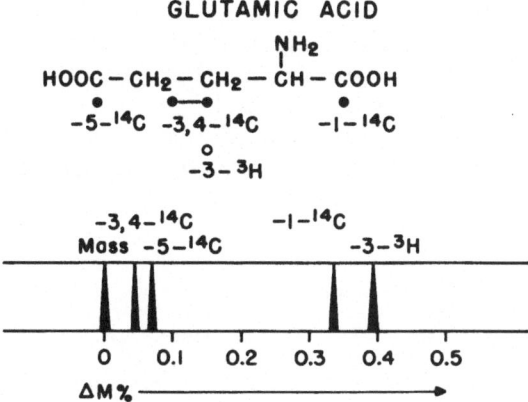

Fig. 4. Isotopic displacement of labeled glutamic acid species from mass as determined by ninhydrin measurement.

atom and the sulfur atom. Since this is not the case observed, a more likely supposition is that the methyl group interacts with the hydrocarbon of the ion-exchange resin during the chromatographic process. If demonstrable, such an interaction would provide more insight into both the ion-exchange process and into functional properties of the methionine molecule in protein chains with respect to the stability of tertiary structures.

These few examples are indicative of the fact that isotopic fractionation effects in analytical systems have moved out of

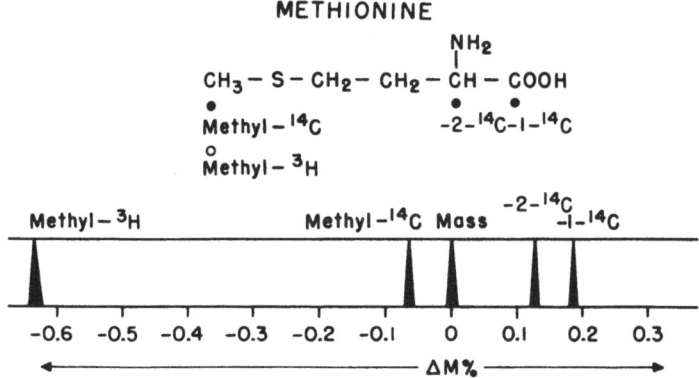

Fig. 5. Isotopic displacement of labeled methionine species from mass.

the realm of laboratory curiosa and have become useful research tools in their own right, enabling us to "see" further into the structure, configuration, and behavior of molecules than ever before. They may be expected to bring to light many new interactions whose energy content is so low as to have evaded detection to date as well as to clarify mechanisms and processes at a more macroscopic level.

REFERENCES

1. Klein, P.D., in: Giddings, J.D., and Keller, R.A. (editors), Advances in Chromatography, Marcel Dekker, New York 3:3 (1966).
2. Laragh, J.H., Sealey, J.E., and Klein, P.D., Proc. Symp. Radiochem. Methods Analysis 2:353 (1965).
3. Klein, P.D., Simborg, D.W., and Kunze-Falkner, B.A., Proceedings of the Second International Symposium on Methods of Preparing and Storing Marked Molecules, Euratom, in press.
4. Klein, P.D., Separation Sci. 1:511 (1966).
5. Cejka, V., Venneman, E.M., Belt-Van den Bosch, N., and Klein, P.D., J. Chromatog. 22:308 (1966).
6. Liberti, A., Cartoni, G., and Bruner, F., J. Chromatog. 12:8 (1963).
7. Ingold, C.K., Raisin, G.G., Wilson, C.L., and Bayley, C.R., J. Chem. Soc. 915 (1936).
8. Saha, N.C., and Sweeley, C.C., Anal. Chem., in press.
9. Klein, P.D., Simborg, D.W., and Szczepanik, P.A., Pure Appl. Chem. 8:357 (1964).
10. Piez, K.A., and Eagle, H., Science 122:968 (1955).
11. Piez, K.A., and Eagle, H., J. Am. Chem. Soc. 78:5285 (1956).
12. Peyser, P., private communication, 1966.

HIGH-RESOLUTION AUTORADIOGRAPHY OF H³-ESTRADIOL WITH UNFIXED, UNEMBEDDED 1.0μ FREEZE-DRIED FROZEN SECTIONS

Walter E. Stumpf and Lloyd J. Roth

Department of Pharmacology
The University of Chicago
Chicago, Illinois

The autoradiographic localization of labeled compounds imposes severe limitations on the selection of procedures to which the tissue is subjected in the process. Unfortunately, this fact has not been fully recognized, and numerous studies of questionable value have been carried out with unsuitable methods. Since precise localization is the purpose of high resolution autoradiography, all steps in the procedure which are likely to cause translocation of the label must be rigidly controlled or excluded. The successful autoradiographic localization of H³-thymidine, which is incorporated into nuclear DNA, has led many people to regard a procedure which is valid in the autoradiography of incorporated thymidine as equally valid in the study of diffusible compounds.

In the development of an autoradiographic method designed to avoid liquid fixation, treatment of the tissue with solvents, and embedding, we reinvestigated the effect of these procedures and found that their use may introduce serious localization artifacts. Such artifacts may remain hidden to the investigator who is misled into a sense of security by methods which provide consistent and reproducible data. While reproducibility is a necessary condition, it is not sufficient to establish validity. Inclusion of one or several steps with a known or potential threat of alteration in tissue constituent-topography imposes the burden of proof on the investigator that no significant alteration has occurred. Interpretation without concern for possible sources of error places the conclusions on a sub-

Supported in part by U.S. Public Health Service Grant NB-04500-04, and American Cancer Society Institutional Grant 41-G Pharmacology.

jective rather than an objective level and invalidates the conclusions derived from the autoradiogram.

FIXATION

As used in histology, fixation is a process by which proteins in the tissue are denatured and, without evidence, should not be construed to mean binding, incorporation, trapping, or immobilization of compounds administered as tracers. That fixation is likely to lead to diffusion and loss of tissue constituents as well as tracer substances has been confirmed. Experimenting with thorium X, Harbers and Neumann [1] have shown that absolute alcohol at $+20°C$ and at $-27°C$ removed 24 and 23% of the radioactivity, respectively, from unfixed liver and 58 and 53% from kidney after 16 hr of fixation. Acetone at $-27°C$ removed 22% of radioactivity from liver and 24% from kidney. Treatment with 10% Formol produced a loss similar to absolute alcohol at $-27°C$. The loss of activity with Carnoy and Bovin solution was 46% in liver. In kidney, Carnoy produced a loss of 66%, while Bovin produced a loss of 25%. Kaminski [2] found with various fixatives that the amount of radioactivity lost during fixation and dehydration varied from 25 to 90% for P^{32} and from 4 to 20% for I^{131}. Hempel [3] injected dl-DOPA [a-C^{14}] intraperitoneally into mice and after 2 days, when all the DOPA had been converted to norepinephrine and epinephrine, removed the suprarenal glands, fixed one in either formalin-calcium chloride or in bichromate solution and prepared the contralateral gland by freeze-drying for control. By scintillation counting he found 90% of the catecholamine activity was extracted by formalin fixation, and 20 to 40% was extracted by bichromate fixation when compared with freeze-dried glands.

Treatment of tissue with fixatives may remove the tracer substance and also has been found to extract significant amounts of tissue as measured by reduction in tissue mass. Merriam [4] reported that 10% formalin fixation extracted 6% of the total protein of liver, 4% from muscle, and 12% of the total protein from brain. In addition, he found detectable amounts of nucleic acids extracted from liver. Arnold and Schneider [5] also found that nucleic acids were extracted in a number of fixation media and that these extracted nucleic acids could be separated by thin-layer chromatography. Sylven [6] fixed tissue slices from perfused rabbit organs in

5% formaldehyde, absolute alcohol, and in Carnoy solution. After 24 hr he found a loss of nitrogen constituents ranging from 4 to 15% in liver, lung, muscle, and spleen. A loss in tissue mass ranging from 10 to 30% calculated on a dry-weight basis was also reported. This investigator concluded that "these defects will invalidate most qualitative and quantitative cytochemical data."

Fixation by perfusion with glutaraldehyde has been recommended because little change was found in the dry weight, electrolyte, and lipid content of the tissue following perfusion [7]. Although glutaraldehyde may be a useful fixative for histological purposes, its use in autoradiography should be questioned since the process of liquid fixation involves the penetration of the tissue by a solvent. It is unlikely that such an infiltration process can occur without concurrent translocation within the tissue of soluble and unbound tracers. The fact that H^3-estradiol could not be demonstrated in rat uterus, which had been fixed by glutaraldehyde followed by embedding in glycolmethacrylate [8], suggests that the H^3-estradiol activity may actually have been leached from the tissue in the process (Table III). Although vapor fixation has been recommended to avoid liquid infiltration of tissue, vapor fixation of freeze-dried tissue with gaseous formaldehyde has also been reported to cause diffusion of catecholamines if the humidity of the gas is not carefully controlled [9].

DEHYDRATION BY FREEZE-SUBSTITUTION

Freeze-substitution was introduced to circumvent graded solvent dehydration and vacuum freeze-drying of tissue, and frozen-section freeze-substitution was designed to circumvent the need for the additional embedding step. Both techniques have been recommended for autoradiography of diffusible compounds. The fact that tissue is exposed to a liquid phase, albeit cold, in a process where lipids and proteins have been extracted [10-12], demonstrates that freeze-substitution with organic solvents cannot be applied with confidence in autoradiography of diffusible compounds.

EMBEDDING

The use of liquid fixatives and dehydration agents has been abandoned in many autoradiographic procedures with the retention of embedding as an innocuous step. The use of embedding, however, has also been questioned [13] as a source of artifact. Edwards, in an autoradiographic study of the cellular localization of Li^6 and B^{10} in brain and brain tumors, found activity leached beyond the tissue margin and concluded that "leaching by the embedding material must be rigidly excluded before conclusive autoradiographic localization can be determined" [14]. The use of resin as embedding media avoids organic solvents commonly used in de-embedding and, therefore, has been recommended for the autoradiography of diffusible compounds. We have demonstrated [15] by scintillation counting of the embedding supernate and by autoradiography, however, that the two compounds H^3-mesobilirubinogen and H^3-estradiol were translocated by embedding with epoxy resin even after vapor fixation of freeze-dried tissue (Tables I and II).

CONFLICTING RESULTS

Previous autoradiographic reports on the localization of H^3-estradiol and H^3-hexestrol in the uterus [8, 16—19] have shown a discordant variety of localizations over glandular cells, in glandular lumen, in apical poles of glandular cells, in cytoplasm, and in nuclei and cytoplasm, and even absence from uterine tissue with activity in erythrocytes only. The dose in these experiments ranged from near physiological to a high

Table I. Activity in Embedding Material by Liquid Scintillation Counting [15] *

Method number	Embedding material	Fixative	H^3-mesobili-rubinogen (rat liver)	H^3-estradiol (rat uterus)
			cmp in supernate	
4	Epoxy†	Formaldehyde vapor	237	32
4	Epoxy‡	Osmium tetroxide vapor	224	42
5	Paraffin	Unfixed		299

*Activity was measured only in epoxy with accelerator (second step). Epoxy without accelerator (first step) was not measured; thus, the total amount diffused into epoxy is probably higher than indicated here.
†Epoxy used for preparation of the tissue for slides 42, 43, and 48 (Table II).
‡Epoxy used for preparation of the tissue for slides 49 and 64 (Table II).

Table II.   Autoradiographic Detection of Activity in Epoxy
Following Embedding [15]*

Slide number	Vapor fixative	Days exposed	Emulsion background	Epoxy in cleft between tissue	Epoxy remote from tissue
42	Formaldehyde	12	3.34 ± 0.23	8.2 ± 0.57	3.7 ± 0.09
43	Formaldehyde	12	3.0 ± 0.39	9.5 ± 0.99	6.6 ± 0.55
48	Formaldehyde	21	1.0 ± 0.12	17.0 ± 0.71	7.97 ± 1.09
49	Osmium tetroxide	21	1.24 ± 0.23	7.07 ± 0.89	6.0 ± 0.34
64	Osmium tetroxide	32	2.1 ± 0.28	18.1 ± 0.41	

*H³-estradiol, rat uterus, method 4. Each slide contained four to six 1-μ sections.
Silvergrains (SG) were counted in 10,000-μ areas, five areas of emulsion, and five
areas of epoxy per slide. SG counts over epoxy in clefts between tissue and at the
margin of tissue were higher than counts over epoxy areas remote from tissue.
The mean SG count and the standard error of the mean are given in grains/1000μ^2.
One batch of emulsion was used for slides 42 and 43, another for slides 48, 49,
and 64. (Reprinted from Stumpf, W.E., and Roth, L.J., J. Histochem. Cytochem.
14:274, 1966.)

dose of 1.1 mg H³-estradiol per mouse.   This work is summarized in Table III.

We found similar inconsistencies in the localization of
H³-estradiol when frozen sections were allowed to melt, when
tissue was embedded or when liquid emulsion was applied.
(1) When frozen sections were mounted on emulsion-coated
slides by thawing, poor resolution resulted with distribution of
radioactivity over both cytoplasm and nuclei; (2) when thawed
sections were dipped in liquid emulsion, a random distribution
resulted with no distinctive association with subcellular
structures; and (3) when freeze-dried vapor fixed tissue was
embedded with paraffin or epoxy resin, translocation of
activity into the embedding media occurred.   In the case of
paraffin, the de-embedded tissue was devoid of activity in the
autoradiogram, while the epoxy-embedded tissue showed a
spreading of activity when compared with the autoradiograms
prepared by the use of freeze-dried frozen sections dry-mounted
on photographic emulsion [15, 21].

HIGH-RESOLUTION AUTORADIOGRAPHY WITH DRY-
MOUNTED, FREEZE-DRIED FROZEN SECTIONS

The diffusion or translocation of the radioactive tracer
within the tissue has been a major obstacle in the development

Table III. Inconsistencies in Autoradiographic Localization of Estrogens in Uterus

Compound Spec. Act. Reference	Dose Route of Application	Animal	Autoradiographic method	Localization (time of excision)
6,7-H³-estradiol 3.85 mc/mg [16]	50 μg/mouse in olive or poppy seed oil; Injected at arbitrary stages of estrus cycle	Mouse	20μ frozen sections spread on gelatin-covered slides and treated with liquid emulsion	Grains in the endometrial cells and in the lumen of the glandular tubes, in contact with the apical poles of the cells (3–4 hr after administration)
6,7-H³-estradiol 150 mc/mg [17]	0.03 μg/ml incubation in culture medium with 0.25% ethanol	Rat	6μ sections fixed in 10% formol saline, paraffin-embedded, mounted with wet stripping film	Gross distribution: "Somewhat variable, some sections showed a fairly even light labeling over all the tissues, but in others the luminal epithelium was labeled more heavily"; cellular distribution: "Cytoplasmic rather than nuclear." (1hr incubation at 37°C)

Compound	Dose	Animal	Method	Findings
3,17-H³-estradiol 0.36 mc/mg [18]	1.1 mg/mouse in peanut oil Subcutaneous	Mouse	5 µ frozen sections freeze-dried and caught on tape, attached to photographic emulsion coated with egg-albumin glycerin	"A specific localization in the cells of the endometrium and in the tubular glands ... nuclei contained more activity than the cytoplasm." (4 hr after administration)
H³-hexestrol 100–150 mc/mg [19]	1–20 µg/kg in ethyloleate Subcutaneous	Cat (ovariectomized)	5–7 µ frozen sections presumed to be thawed on glass slide, then dipped in bulk emulsion according to previous publication [17]	"Grains overlie the apical poles of gland cells subjacent to the gland lumen." (4 hr after administration)
6,7-H³-estradiol Spec. act. unreported [8]	0.73 µg/rat 7.3 µg/rat in saline with 5% ethanol Subcutaneous	Rat (ovariectomized and non-ovariectomized)	0.2–2 µ sections 3 diff. fixation embed. proced.: 1) glutaraldehyde-paraffin...... 2) osm. acid-methyl butyl methac.. 3) glutarald.-glycol methacryl... All dipped in liquid emulsion	occasional at apex of epith. cells .. apex & base of epith. cells & rbc no satisfactory results (1 hr after administration)

of autoradiographic techniques for diffusible compounds. After reviewing the various procedures employed to solve this problem, we concluded that classical histological techniques, such as liquid fixation, graded solvent dehydration, embedding, and wet-mounting, would probably have to be abandoned in high-resolution autoradiography of diffusible compounds since each carries the implicit threat of translocation.

The method we have developed consists of the following steps:

1. Freezing of tissue in liquefied propane at -180°C with simultaneous mounting on tissue holder for cryocutting.
2. Cryocutting of tissue at 1.0μ or thinner performed at knife temperatures at or below -60°C.
3. Vacuum freeze-drying of thin, frozen sections utilizing a portable cryosorption pump (Fig. 1).
4. Dry-mounting, preferably in a low-humidity dark room, of unfixed, unembedded freeze-dried sections on emulsion-coated slides (Kodak NTB-3) previously stored over Drierite.
5. Exposure in a freezer at -15°C.
6. Photographic development with staining as the final step.

RESULTS AND DISCUSSION

Using the method described above, the subcellular distribution of $6,7-H^3-17\beta$-estradiol (specific activity 208 $\mu c/\mu g$) was studied in uteri of mature ovariectomized and immature Sprague Dawley rats with physiological and near physiological doses ranging from 0.082 to 0.63 μg per animal [22]. These experiments were carried out in collaboration with E. V. Jensen and P. W. Jungblut of the Ben May Laboratory for Cancer Research, University of Chicago. The animals were sacrificed by decapitation at 15 min, 2 hr, and 6 hr after subcutaneous injection of the labeled hormone dissolved in 0.5 ml saline. Tissue pieces approximately 1 to 2 mm^3 were excised and frozen without delay in liquefied propane and sectioned in the cryostat and autoradiograms were prepared in the manner described above. The radioactive purity of the H^3-estradiol was determined by paper chromatography using the Bush III system. Extraction of tissue revealed no metabolites [23].

Using the method of dry-mounted, freeze-dried frozen

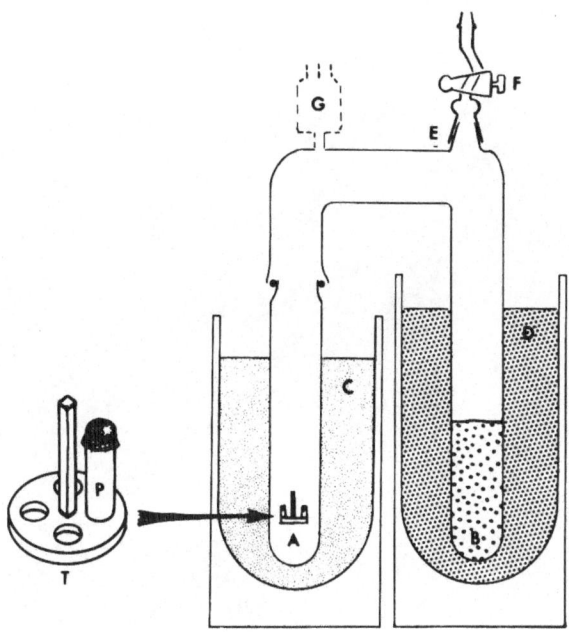

Fig. 1. Portable freeze-drying unit (Delmar Scientific Laboratories, Inc., Maywood, Illinois) is charged through E with 200 g of molecular sieve (Linde 5A-30, Pellets 1/16 in.) in finger B. Frozen sections are cut at 1.0 μ in the cryostat (International Equipment Company, Needham Heights, Mass,), transferred to polyvial P without rise in temperature and then transferred in carrier T to specimen finger A, which has been precooled in the cryostat at -85°C. The freeze-drying unit is assembled in the cryostat and evacuated while in the cryostat through F (vacuum cup stopcock) with a mechanical pump to approximately 10^{-2} mm Hg. F is closed, the unit is removed from the cryostat, and fingers A and B are immersed simultaneously in alcohol dry ice slush (C) and liquid nitrogen (D), respectively. The vacuum is measured by an ion gauge G with an Ion Gauge Control (NRC Equipment Company, Newton, Massachusetts). Vacuum of at least 1.0×10^{-5} mm Hg is readily maintained. Drying of frozen sections is usually completed in less than 12 hr [21].

sections, H^3-estradiol was found to be localized nearly exclusively in nuclei of epithelial cells, cells of the lamina propia, and muscle cells from uterine tissue. Radioactivity was also found to be localized over red blood cells and in adventitial cells of the uterine vasculature. The intensity of labeling varied between the individual cells. Few cells contained no activity, while others were heavily labeled. Most of the cells contained an intermediate amount of activity. In each case, nuclear localization predominated, with no distinction in the general degree of nuclear labeling existing between

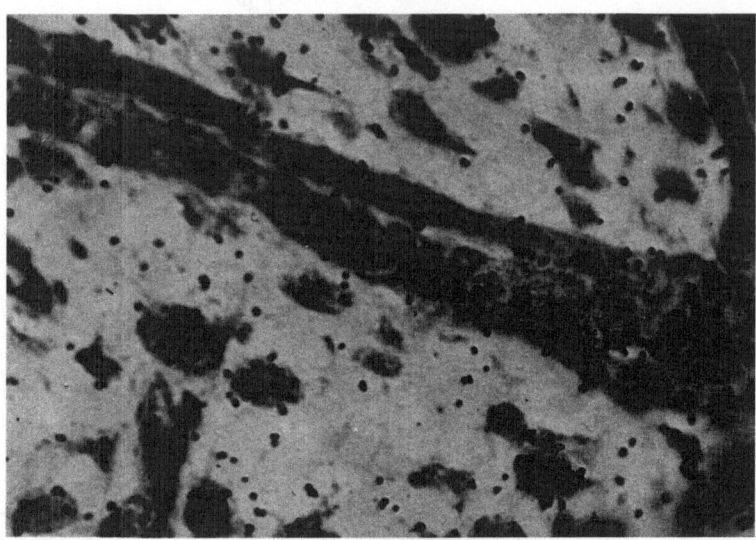

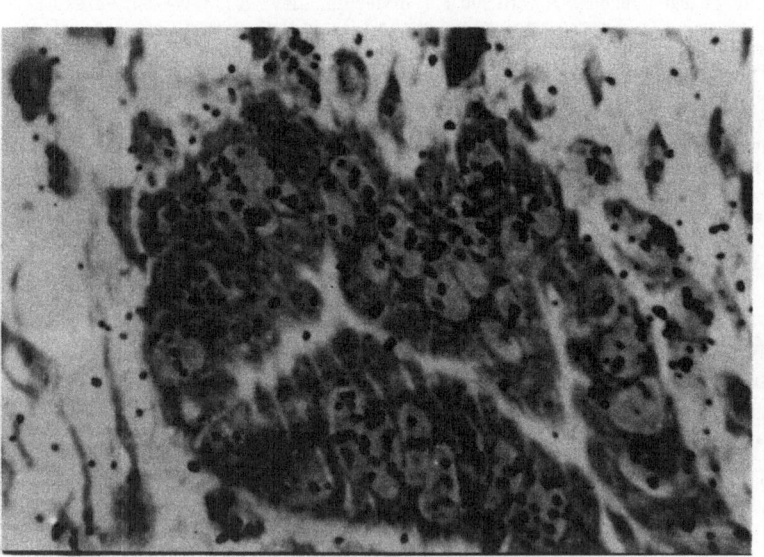

Fig. 2. (A) Autoradiogram of uterus from 24-day-old rat, 2 hr after subcutaneous injection of 0.63 μg H³-estradiol in 0.5 ml saline. Concentration of radioactivity in cell nuclei of uterine glands and lamina propia. One-micron sections stained after photographic development for DNA and RNA with methyl-green-pyronin, ×1200. Exposure 42 days. Photographed through Zeiss didymium and fluorescence barrier filter 65. Type C prints made from 35-mm transparencies through an internegative. Fig. 2 (B) Same autoradiogram as in Fig. 2 (A) but photographed through Zeiss didymium and fluorescence barrier filters 50 and 65 using Kodachrome II professional film. Exposure 51 days at −15°C.

epithelial cells and cells of the lamina propia. Muscle cells, however, appeared to be somewhat less heavily labeled than endometrial cells. No significant difference was noted with the various conditions of time following injection or dose and age of the animal used (Fig. 2).

These findings are in contrast with those compiled in Table I with respect to both the localization in the various cell types and the subcellular localization. We have not seen localization in glandular lumen nor at apical or basal poles of uterine cells as reported.

H^3-Hexestrol, a synthetic estrogen, was included in Table III since it has been shown to resemble estradiol in its affinity for target tissues [23] and to have a physiological response similar to estradiol in producing behavioral estrus. The reported localization of hexestrol in the apical pole of endometrial gland cells in ovariectomized cat, using thawed frozen sections, followed by dipping in liquid emulsion, probably results from translocation from the nucleus in the thawing of the 5 to 7 μ sections and the subsequent application of wet photographic emulsion. A nuclear localization of hexestrol could be expected similar to estradiol, although there remains the possibility that its true localization is the apical pole and different from that of estradiol since it is a compound with a different structure. This is an important question which may be resolved by the use of dry mounting of freeze-dried frozen sections.

In contrast to the localization of estrogens, H^3-norethynodrel, a synthetic progestin, was found to be extranuclear in rat uterine tissue (unpublished results) by the method we describe. This nonnuclear localization in uterine tissue also argues against the possibility that the nuclear localization of estradiol is the result of a chemographic artifact related specifically to the nuclei.

Confidence in the method we report is based on (1) the exclusion of liquid fixation, solvent dehydration, embedding, and de-embedding of tissue, (2) on freeze drying at $-68°C$ with avoidance of thawing by premature rise in temperature [21], (3) on the use of thin tissue sections which reduce superimposition of structure and provide high resolving power, and (4) on the use of emulsion-coated slides, previously dried over Drierite, with mounting of the freeze-dried sections at room temperature or in a cold room at $-8°C$ with photographic exposure in a deep freeze at $-15°C$.

CONCLUSIONS

High-resolution autoradiography is a sensitive tool for the subcellular localization of labeled compounds which requires the rigorous exclusion of any and all steps which may result in translocation of the label in the tissue. Any step in the autoradiography of diffusible compounds in which substantial loss of radioactivity occurs is generally recognized as an invalidating step. The loss of small amounts of radioactivity to solvents or embedding media in an autoradiographic procedure may seem reassuring, but in fact it is alarming since such loss argues for redistribution of the label within the tissue.

Although the binding of the tracer by tissue has been invoked to provide assurance that treatment of the tissue with fixatives and solvents is without risk, the extraction of protein [4–6] and lipids [4, 5, 12, 24] suggests caution in the use of this argument. The degree of binding may be altered not only in such a procedure but in some cases the binding complex may actually be removed from the tissue. Furthermore, false binding of amino acids to tissue induced by fixation has also been reported [25, 26].

Peters and Ashley [25] found that autoradiograms prepared after fixation with glutaraldehyde, osmic acid, or formaldehyde revealed that 63, 25, and 4%, respectively, of the silver grains were due to fixation-induced binding of free amino acids.

Since justification of the results and their confirmation depends on the methodology, it is essential that the steps employed be described in detail. Too frequently autoradiographic results have been taken for granted without sufficient concern for the possible sources of error which may have been introduced.

REFERENCES

1. Harbers, E., and Neumann, K., Klin. Wschr. 32:337 (1954).
2. Kaminski, E. J., Stain Technol. 30:139 (1955).
3. Hempel, K., Histochemie 4:507 (1965).
4. Merriam, R.W., J. Histochem. Cytochem. 5:43 (1957).
5. Arnold, M., and Schneider, R., Histochemie 7:260 (1966).
6. Sylven, B., Intern. Union Against Cancer 7:708 (1951).
7. Levy, W.A., Herzog, I., Suzuki, K., Katzman, R., and Scheinberg, L., J. Cell Biol. 27:119 (1965).
8. Inman, D.R., Banfield, R.E.W., and King, R.J.B., J. Endocrinol. 32:17 (1965).
9. Falck, B., and Owman, C., Acta Univ. Lund., Sect. II, No. 7:5 (1965).

10. Chang, J.P., and Hori, S.H., J. Histochem. Cytochem. 9:292 (1961).
11. Chang, J.P., and Hori, S.H., Ann. Histochem. 6:419 (1962).
12. Neumann, K., in: Graumann, W., and Neumann, K. (editors), Handbuch der Histochemie, Vol. 1, Gustav Fischer Verlag, Stuttgart, 1958.
13. Harbers, E., in: Graumann, W., and Neumann, K. (editors), Handbuch der Histochemie, Vol. 1, Gustav Fischer Verlag, Stuttgart, 1958.
14. Edwards, L.C., Stain Technol. 30:163 (1955).
15. Stumpf, W.E., and Roth, L.J., J. Histochem. Cytochem. 14:274 (1966).
16. DePaepe, J.C., Nature 185:264 (1960).
17. Mobbs, B.G., J. Endocrinol. 27:129 (1963).
18. Ullberg, S., and Bengtson, G., Acta Endocrinol., Copenhagen 43:75 (1963).
19. Michael, R.P., Brit. Med. Bull. 21:87 (1965).
20. Glascock, R.F., and Michael, R.P., J. Physiol., Lond. 163:38P (1962).
21. Stumpf, W.E., and Roth, L.J., J. Histochem. Cytochem. 15:243 (1967).
22. Stumpf, W.E., and Roth, L.J., Pharmacologist 8:212 (1966).
23. Jensen, E.V., Jacobson, H.I., Flesher, J.W., Saha, N.N., Gupta, G.N., Smith, S., Colucci, V., Shiplicoff, D., Neumann, H.G., DeSombre, E.R., and Jungblut, P.W., in: Pincus, G., Nakao, T., and Tait, J.F. (editors), Steroid Dynamics, Academic Press, New York, 1966.
24. Morgan, T.E., and Huber, G.L., J. Cell Biol. 32:757 (1967).
25. Peters, T., and Ashley, C.A., J. Cell Biol. 33:53 (1957).
26. Antoni, F., Koteles, G.J., Hempel, K., and Maurer, W., Histochemie 5:210 (1965).

RADIOAUTOGRAPHY AS A TECHNIQUE IN TRACER METHODOLOGY

N. J. Nadler

Department of Anatomy
McGill University
Montreal, Canada

Radioautography is a technique to detect by means of photographic emulsions the site of radioactive elements in tissue sections. Accordingly, in biology, radioautography is an ideal tool to trace the course of substances in and out of various microscopic or ultramicroscopic structural entities [1].

It is of interest that, just before the turn of the century, it was the accidental recording on a photographic plate which led Becquerel to the discovery of radioactivity. In the first quarter of the century, photographic plates were used to record the presence of radioactive radium and other heavy elements, but it was only about 1940 that the technique of radioautography was taken up in earnest dealing with substances of real biological importance. At that time the method consisted of placing a tissue section, in which a radioactive element was contained, in apposition with a photographic plate. This yielded results which were crude according to today's standards. For example, in 1940 it was possible to detect a minimum of 100 or so radioactive emissions and to localize the site of radioactivity in the tissue to within around $100\,\mu$. Now, using advanced methods, it is possible to pick up just one emission and to localize the source of the single radioactive atom to less than $0.1\,\mu$ in the tissue.

Radioautography has been used with considerable success to investigate the turnover of metabolic substances in many organs. My own experience being mainly with the thyroid, I intend to use this gland as a model to illustrate what can be achieved with this technique.

As an interjection about the structure of the thyroid, the gland is made up of several hundred thousand individual

127

follicles, each about 150μ in diameter. A follicle works as an independent unit, so that if we wish to know how the whole gland works we need to focus only on any follicle. Each follicle consists of a single layer of epithelial cells enclosing a central lumen which contains a viscous homogeneous fluid called colloid.

In the thyroid follicles, inorganic iodide is taken up and is bound to a protein called thyroglobulin. Iodide is attached to the tyrosyl radicals of the thyroglobulin molecules, and within the framework of these protein molecules the iodotyrosyls combine to form iodothyronyls. Then, still in the follicles of the gland, the iodinated protein molecules are proteolyzed, and the iodothyronyls are liberated as free 3,5,3'-triiodothyronine or free 3,5,3',5'-tetraiodothyronine. These amino acids, relatively small molecules, escape the thyroid to go into the circulation as thyroid hormones.

Now let us examine in detail the synthesis within a follicle of the protein moiety of thyroglobulin. Because leucine happens to be the most abundant amino acid in the thyroglobulin molecule, leucine labeled with tritium was selected to be injected into rats. When the thyroid gland was removed very shortly after injection and a radioautograph prepared, the site of the leucine label demonstrated the location of incorporation of radioactive leucine precursor into newly synthesized thyroglobulin. In conventional histological preparations, all water-soluble substances, in fact all relatively small molecules, are washed out, and what we have are only the fixed heavy molecules such as proteins. (To demonstrate the presence of the smaller molecules, the technique must be modified as discussed by another speaker in this symposium.) Thus, after histological processing, the tissue sections are coated with a photographic emulsion by dipping the slides into melted emulsion. The preparation is then allowed to expose. During this time, β-rays emitted from the sources of radioactivity (H^3-leucine bound in protein) will strike the overlying photographic emulsion and affect the silver bromide crystals in the emulsion. Those crystals so affected, close to the source of radioactivity in the tissue, after photographic development will show up under the light microscope as black silver grains.

In the experiments conducted, at 10 min after H^3-leucine injection, the location of the silver grains showed us that the label was incorporated in protein present in the cytoplasm of all cells in all follicles. We know that within 10 min after injection of H^3-leucine, newly labeled thyroglobulin molecules could not have moved to any extent from where they are synthesized; thus the site of synthesis of the protein moiety of thyroglobulin must take place in every follicle in the cytoplasm of the cells. By 4 hr after injection, radioautographs showed that silver grains are concentrated as a band at the apex of the follicle cells bordering the colloid. This means, then, that after synthesis newly formed thyroglobulin molecules in the cells migrate in the direction of the colloid. After 4 hr, radioautographs revealed silver grains over the colloid, meaning that the label is in the colloid and that the newly synthesized thyroglobulin must have been secreted by the cells into the colloid of every follicle.

Thus far we have been concerned only with resolution in radioautography, that is, to localize the site of radioactive substances in tissues. However, radioautography can also be used for quantitative purposes by counting silver grains over selected regions. Of course, it is necessary to exercise care to keep well controlled all factors which affect the intensity of the reaction, such as section thickness, interspace between section and emulsion, development conditions, and so on. Nevertheless, by grain counting it is possible to measure the relative content of radioelement in various structures at various times. Thus, it was shown that in the thyroid the protein label concentration appeared to rise to a maximum first in the base of follicle cells, then in the apex of the cells, later in the periphery of the colloids bordering the cells, and finally throughout the whole of the colloid—indicative of the synthesis of the protein moiety of thyroglobulin in the cells, of its migration to the apical region of the cells, and of its secretion into the colloid. From quantitative data of this kind it was possible to derive conclusions on the rate of synthesis of thyroglobulin and of the rate of its secretion by cells in the thyroid follicles, correlating these rates with the volume of the responsible morphologic entities—the correlation of function and structure.

More recently this phenomenon was investigated closely by means of electron-microscope radioautography. In this procedure a radioautograph is made by coating a section of thickness 0.05μ with a monolayer of silver bromide crystals in a photographic emulsion. The mount is allowed to expose and is then developed according to photography. Finally the whole combination is examined in the electron microscope, where the silver grains look like lacy black bits.

Early, 10 min after injection of H^3-leucine, silver grains were discovered over the ribosomes in the endoplasmic reticulum, depicting this as the site of synthesis of thyroglobulin. Later the radioautographic reaction is over the Golgi zone, showing that newly synthesized thyroglobulin protein molecules have migrated there. According to the radioautographic results at serial times after H^3-leucine injection, from the Golgi, packaged in membrane-enclosed vesicles, the labeled thyroglobulin molecules migrate to the apical surface of the follicle cells and are then released into the colloid.

After the protein moiety of thyroglobulin reaches the colloid in the thyroid follicle, iodination takes place there. Proof of this came from investigations using radioautography after injection of I^{125}. Almost immediately after administration, conventional radioautographs show that the iodine label is present in the colloid. In special freeze-dried radioautographs (unpublished with B. Benard and C. P. Leblond) designed to demonstrate diffusible substances, it was possible to show that inorganic iodide was taken up by the follicle cells and deposited into the colloid. So there in the colloid, iodide is bound to the protein moiety of thyroglobulin to yield the fully mature iodinated molecule. These phenomena, the uptake of iodide, and its binding to protein have also been the object of quantitative investigations; by grain counting, it has been possible to correlate the rate of iodine metabolism in any follicle with the volume of follicle cells contained therein.

To detect the last stage in the formation of thyroid hormone, I^{125} was injected to label thyroglobulin in the colloids of the follicles; then thyroid stimulating hormone (from the anterior pituitary) was injected to accelerate the process of thyroid hormone release. Radioautographs depicted what happens: The follicle cells send out streamers of cytoplasm and pick up bits of colloid (pinocytosis), bringing iodinated thyroglobulin protein molecules back into the cells as intracellular colloid droplets. We are certain that the protein molecules are bound with iodine because there were silver grains over the intracellular colloid droplets. By the action of intracellular enzymes, the iodinated thyroglobulin molecules are proteolyzed to release the free iodothyronines which leave the cells to go into the circulation as thyroid hormones.

In summary, radioautography is a technique which lends itself nicely to the examination of the turnover of substances in tissues. The method can be used to trace the passage of a labeled substance through various microscopic or ultra-microscopic structural entities and, by quantitative applications, yield data relating to the kinetics of the turnover.

REFERENCE

1. General references to the technique of radioautography are compiled in Symposia of the International Society for Cell Biology, The Use of Radioautography in Investigating Protein Synthesis, Academic Press, New York, 1965. Specific references to the thyroid gland are contained in the chapter by N. J. Nadler, "The Elaboration of Thyroglobulin by the Thyroid Follicular Cells."

LIQUID SCINTILLATION COUNTING: AUTOMATED MATHEMATICAL FITTING AND USE OF CHANNELS RATIO METHODS BY COMPUTER PROGRAM

F. A. Blanchard, Mary R. Wagner, and I.T. Takahashi

Radiochemistry Research Laboratory
The Dow Chemical Company
Midland, Michigan

INTRODUCTION

One way of expressing the liquid scintillation counting job is simply that after a tracer experiment yields some samples one wants to know how much tracer is in each. These samples may vary greatly in physical, chemical, and color properties. These properties affect the scintillation process. If we rely on relative count rates for comparing samples, we may be greatly deceived.

Converting all cpm to absolute dpm values as a common basis is generally recognized as highly desirable. Witness the variety of standardizing techniques which have been developed. Review articles by Peng [1] and by Frank [2] discuss some of the advantages and disadvantages of such methods as internal standardization, channels ratio, and external source.

All of these require a great deal of calculating. For example, with channels ratio one must prepare a plot from standards. From this an efficiency must be read for each sample. This is then used for an additional division.

Counting instrument design has recognized this job. Background subtract and ratio calculators are a step in this direction. Conversion all the way to dpm is claimed by one new counting instrument [12].

Instead of trying to include in the hardware of counting instruments all the changing and various calculating needs, one can take the basic time and counting data directly from the liquid scintillation counter and enter it into a computer. This can be programmed for efficiency calculations by any of the methods. The program can also provide for a variety of other data manipulation. The combination of punched data output and computer processing can give an operation which is fast, accurate, flexible, and automatic.

In 1963 we reported on our use of computers for this [3]. An ALGOL 58 program handled the reading-in of counting data. It sorted the data, converted counts to cpm, tested them for compliance with radioactivity statistics, averaged them over repeat cycles, corrected for background, derived and used an efficiency from a fixed value, or a representative or internal standard, and applied multiplying or dividing factors. This program has had steady use. W. R. Koehler, at Lever Brothers, prepared a Fortran modification of this for their use [4].

J.L. Spratt in 1965 reported a Fortran IV program which handles intermixed C^{14} and H^3 samples with three-channel data [5]. Information on which isotope is next is given by skipped sample positions. Efficiency is by internal standardization.

L.R. Axelrod et. al. in 1965 published a Fortran program with emphasis on statistical reliability of the data [6]. To save computer memory it uses a presort of the data cards on a card sorter. It handles quenching by a least-squares fit to a linear equation. Dual label samples are determined by using an external source, with data fitted to three linear equations by least squares.

C. Matthijssen has a program for dual label samples by dual internal standards [7]. Nuclear-Chicago advertises a program to be used in conjunction with their Data Converter Controller which can make use of a channels ratio method. A. Rossel has a program which handles quenching by bracketing the ratio of an unknown between those of standards [11]. It then uses a linear interpolation. There are other more specialized programs. Some of these have been collected and made available through the Packard Instrument Company.

EXAMPLE OF COMPUTER METHOD WITH OUR CURRENT
PROGRAM   .

As an example of what can be done we will describe in some detail the use of a new program we have prepared in ALGOL 60 for use with a Burroughs B 5500 computer. It includes most of the functions of our previous one; in addition, it can work with either two or three counting channels from either punched tape or cards.

The data was obtained with a Tri-Carb® Model 3324 connected to a Friden Add-Punch®, Model ACPTS. Standard Packard names are used for the first, second, and third counting channels: red, green, and blue.

Although a basic goal is to condense a series of time and count values for a sample to a single radioassay in some convenient units, we find it good practice to get a print back of the original data for later reference. This is shown in the top third of Fig. 1. At the top are two required constants: an arbitrary file number and the dpm in the standards. An optional divisor and a multiplier for the final result for each sample would appear under DVSR and MULT if used. Below these are the original counting data arranged by sample number with the three repeat cycles of this run brought together for each sample.

There need be little punched data other than the counting data itself since information as to number of samples, cycles, channels, what ratio to calculate, and location of background samples and of standard and unknown groups can all be derived directly from the counting data. There is compensation for such "quirks" in data as missing counts, punching errors, and erroneous sample numbers before these counts are averaged. This is important when the people doing the counting work are not the ones operating the computer.

Determination of counting efficiency is a key step in the data analysis. We are using the channels ratio method to illustrate this, although the program includes options for representative standards, internal standards, and external source-ratio methods. The channels ratio method itself was set forth by Baillie [8] and refined by Bruno and Christian [9], Bush [10], and many others.

We use a variation which might be called an equation method. Each counting run contains a series of quenched standards. An equation relating efficiency to channels ratio is derived by the computer from the counting data of these standards. This is then used with subsequent unknowns.

The program uses a flexible but simple arrangement of samples in the counting sequence. Sample one is used as a background. Succeeding samples up to a skipped sample number are standards. Any number more than four may be used. The center third of Fig. 1 shows how the results from the quenched standards are given. At the top the columns from

```
                      LIQUID SCINTILLATION RADIOASSAY
          FILE      DPM     DVSR      MULT.
          641     38300


  1 TIME      100      100      100
    RED        41       27       31
    GREEN      31       28       30


  2 TIME      100      100      100
    RED      31540    31386    31398
    GREEN     7588     7663     7720


  3 TIME      100      100      100
    RED      28014    28056    27912
    GREEN    11730    11640    11804

                    LIQUID SCINTILLATION RADIOASSAY
                    COUNTING EFFICIENCY STANDARDS :
                           CHANNELS RATIO

    NO    OK   RNET    OK   GNET    G/R  EFFIC       DPM
          R             G

     1    3     33.0   3     29.7  0.90  0.00000  0.000@+00
     2    3  31408.3   3   7627.3  0.24  0.82006  3.830@+04
     3    3  27961.0   3  11695.0  0.42  0.73005  3.830@+04
     4    3  21384.7   3  17463.7  0.82  0.55835  3.830@+04
     5    2  14693.0   3  16995.3  1.16  0.38363  3.830@+04


  A   9.82366@-01    B  -7.86850@-01    C   5.55776@-01    D  -2.79265@-01
  MULT-DETM   1.00000@+00

    R      0.000   E      0.982
    R      0.100   E      0.909
    R      0.200   E      0.845
    R      0.300   E      0.789
    R      0.400   E      0.739
                    LIQUID SCINTILLATION RADIOASSAY
                    COUNTING EFFICIENCY STANDARDS :
                           CHANNELS RATIO

    NO    OK   RNET    OK   GNET    G/R  EFFIC       DPM
          R             G

     7    3  11317.7   3   3679.3  0.33  0.77571  1.459@+04
     8    3  11316.0   3   3728.7  0.33  0.77345  1.463@+04
     9    3  11208.3   3   3426.0  0.31  0.78581  1.426@+04

   AVG    =   1.44947@+04       STD DEV = 2.01275@+02


    11    3  16078.7   3   3582.0  0.22  0.83157  1.934@+04
    12    3  15508.7   3   3509.0  0.23  0.82955  1.870@+04
    13    3  15959.7   3   3511.3  0.22  0.83318  1.916@+04

   AVG    =   1.90619@+04       STD DEV = 3.30080@+02
```

Fig. 1. Liquid scintillation radioassay. Portions of three pages of computer output.

left to right give the sample number, the number of statistically acceptable red counts (± 3 standard deviation of median and average; std dev = $\sqrt{\text{cpm/time}}$); net cpm in that channel (except No. 1 is the actual BG) and similarly for the green channel. G/R indicates the ratio: net green cpm. For these standards the counting efficiency is obtained by dividing the net red cpm by the dpm constant from the beginning of the data set. For each of these, the dpm is by definition the dpm constant. The constants A, B, C, and D define a least-squares fitted reference curve relating efficiency to channels ratio. It is the cubic: $E = A + BR + CR^2 + DR^3$. An index of multiple determination is given to indicate how well the data of the standards fit the equation. Calculated points from the equation are given for use in hand-plotting when one wishes to visualize the curve.

The lower third of Fig. 1 shows results for unknowns. Here the efficiencies are derived from the equation. The counts and the efficiencies are used to get the dpm values. Groupings of unknowns were set up by simply skipping a sample position on the sample belt of the spectrometer. After each such group the average and standard deviation of the dpm values in that group are printed out. This is helpful when the group consists of replicates.

TESTS OF CHANNELS RATIO EQUATION METHOD

How well does this equation method work with the channels ratio method? What is the best way to set it up? We have made many trial runs to try to find out. Most of the testing has been done using sets of five to nine chloroform or methyl red standards to generate the equations. Replicates of these standards as well as other samples with a variety of other chemical and color quenchers were prepared as "unknowns." Groups of five samples at each level of quenching were used. To save time and still have adequate counting statistics, we used about 40,000 dpm of C^{14} or 200,000 dpm of H^3. These were given three to five 1-min counts. The counting efficiencies of the unknowns were determined directly and by the equation method.

Figure 2 shows the quality of fit of computer-generated quadratic and cubic equations as compared to the actual points from chloroform-quenched C^{14} standards. Here the cubic equation is obviously the better fit. This was generally true.

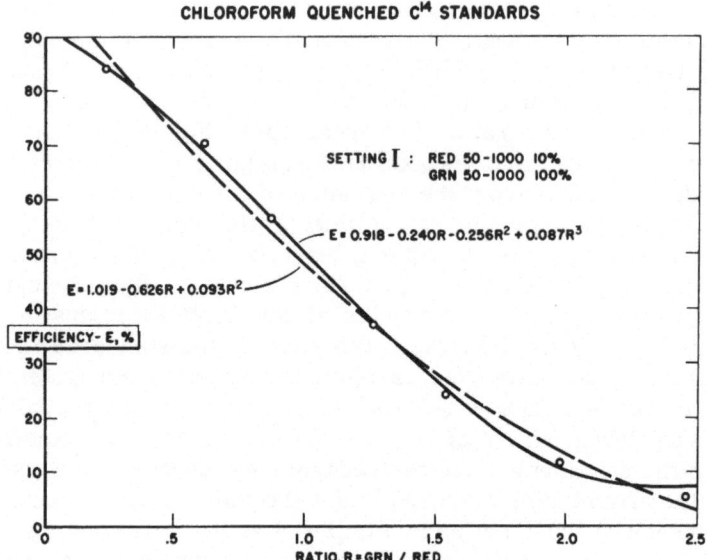

Fig. 2. Chloroform-quenched C^{14} standards.

Subsequent data is all by cubic equations. This particular instrument setting is one we have worked out. We have not seen it reported elsewhere. It is about equivalent to being able to set the green channel at 10% gain with a 5 to 100 window. It spreads out the ratios at low counting efficiencies.

Figure 3 shows similar results for color quenching with a different type of setting. Type 1 setting works equally well. Separate curves are needed for chemical and color quenching for efficiencies lower than about 65%. Figure 4 shows a very good fit with tritium data.

Numerical comparisons of the ability of the equation method to determine the efficiencies of "unknowns" are given in Tables I and II for C^{14} and Tables III and IV for H^3. For C^{14} these can be obtained within three absolute efficiency units over the range 10 to 70% and for tritium, to within 1 unit from 1 to 28%. This assumes that the type of quenching in the standards has been reasonably well matched with that in the unknowns.

EXTERNAL SOURCE RATIO METHOD

This can be used without special source handling and positioning equipment. The procedure is surprisingly simple.

After a normal counting run, a source (for example, 2 mc Cs^{137}) is placed on top of the freezer lid directly over the sample well. Samples are recounted in the presence of this γ-background. Channels ratio calculations are then based on increased counts due to γ's. This works! In preliminary tests, this method appears to be as good or better than the channels ratio method, especially when the E in the equation is replaced with $\ln E$.

OPTIONAL FACTORS AND SUPPLEMENTARY DATA
MANIPULATION

Other calculations easily and routinely available include multiplying or dividing factors for groups of samples or individual samples. A supplementary output is available which can provide averaging by small groupings of 2, 3, and so on. It can also convert these averages to fractions of the sum of a series of these (for example, for chromatograms). It can give each as a fraction of the first of a series (for example, for change in solutions due to adsorption, reaction, and so forth).

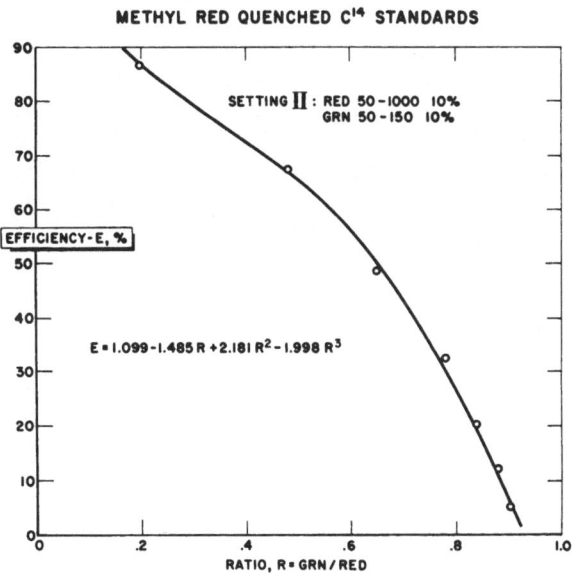

Fig. 3. Methyl red quenched C^{14} standards.

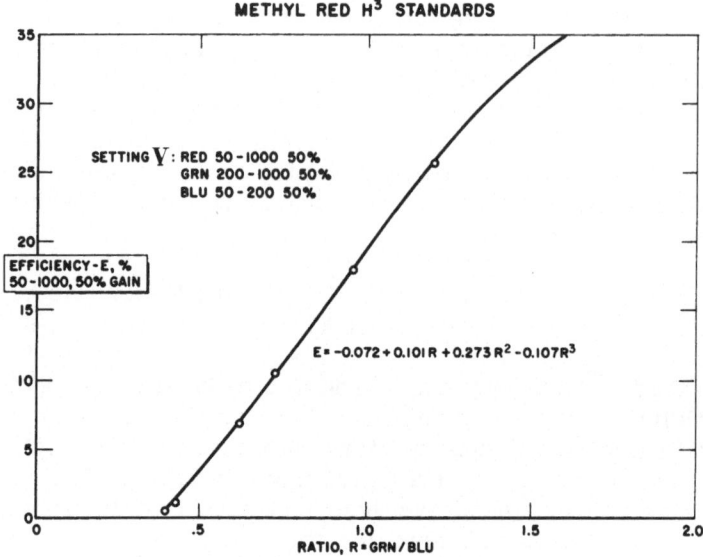

Fig. 4. Methyl red H^3 standards.

CONCLUSIONS

Computer processing can be an effective tool for handling liquid scintillation counting data. The data is readily presented to the computer. It may either be punched on cards or tape during the counting instrument readout or keypunched from the printed tape. Counts can be converted to cpm. These can be checked for errors, averaged, and corrected for background. The channels ratio can be computed. Either the representative standard, channels ratio, internal standard, or external-source-ratio efficiency method can be chosen to convert cpm to dpm. Factors can be applied to the dpm values of single samples or groups of them to convert the radioassay to such units as micrograms or ppm. All of this can be done with a program which is easy to use. Ours calls for only two constants and a single control number to go with the counting data itself. A simple but flexible sample arrangement is used.

Handling the channels ratio method by equations eliminates one stumbling block to its convenient use—the necessity of preparing and using appropriate graphs. By inclusion of care-

Table I.   Accuracy of C^{14} Efficiency Calculation: Chloroform Quenching

Setting		Standards		Test samples	
		E	ΔE	ΔE^*	σE
I: R 50−1000 10%		70	−1.2	−1.76	1.26
G 50−1000 100%		50	+0.5	−0.60	0.96
Ratio G/R		30	+1.0	+1.05	1.14
		10	−1.0	−2.09	0.63
II: R 50−1000 10%		70	−2.0	−2.96	0.48
G 50−150 10%		50	0.0	+1.89	1.31
Ratio G/R		30	−2.2	−4.33	1.11
		10	−0.5	+0.43	0.33

$^*\Delta E = E_{calculated} - E_{actual}$

Table II.   Accuracy of Calculated C^{14} Efficiencies: Several Chemical and Color Quenching Agents*

Standards	E	ΔE		
		Chloroform	Carbon disulfide	Nitro-benzene
Chloroform	70	−1.59	−2.29	−2.10
	50	+0.01	−1.73	+0.89
	30	+0.93	+0.62	+1.23
	10	−2.26	−2.28	−2.14
		Methyl red	Oil red dye	Oil blue
Methyl red	70	−3.25	−2.88	−4.68
	50	+0.34	−2.59	−6.31
	30	+0.14	+0.21	−6.37
	10	−1.28	−2.02	−5.41

*Setting I: R 50−1000 10%
         G 50−1000 100%

Table III.   Accuracy of H^3 Efficiency Calculations:
Chloroform Quenching

E	ΔE				
	Setting*				
	I	II	III	IV	V
28	0.00	−0.21	+0.01	−0.24	−0.14
22	+0.24	+0.61	+0.45	+0.52	+0.78
19	+0.40	+0.71	−0.35	+0.64	+0.73
14	+0.11	+0.15	+0.31	+0.13	+0.03
8	+0.17	−0.23	+0.04	−0.31	−0.14
1.6	−0.18	−0.22	−0.59	−0.10	+0.31
0.9	+0.07	−0.25	−0.68	−0.02	+0.25
0.5	+0.25	−0.17	−0.79	−0.29	+0.17

*Settings:	R	G	B	Ratio
I	50–1000	50–1000		G/R
II	50–1000	50–300		G/R
III	50–1000	300–1000		G/R
IV	50–1000	50–200	200–1000	G/B
V	50–1000	200–1000	50–200	G/B

All gains 50% except in setting I: $R = 30\%$ and $G = 100\%$.

Table IV.   Accuracy of Calculated H^3 Counting Efficiencies:
Several Chemical and Color Quenching Agents*

Standards	E	ΔE		
			Carbon	Nitro-
Chloroform		Chloroform	disulfide	benzene
	25	0.06	0.17	1.20
	13	−0.60	−0.45	−0.15
	6	−0.55		−0.38
	2	0.17	0.22	−0.04
Methyl red		Methyl red	Oil red dye	Oil blue
	26	0.34	−0.76	0.70
	17	0.46	0.90	−0.88
	10	−0.10	0.88	−1.03
	3	0.19	0.19	−0.27

*Setting IV: R 50−1000 ⎫
     G 50−200 ⎬ 50%
     B 200−1000 ⎪
     Ratio: G/B ⎭

fully prepared, variably quenched standards at the beginning of each counting run, the most valid cubic equation for that run is obtained. We have found the accuracies attainable with several types of channels ratio settings about equivalent. C^{14} efficiency can be determined to within three absolute efficiency percentage units over an efficiency range of 10 to 70%. H^3 can be done to within one unit over a range of 1 to 28%. If there is marked quenching, the type of quencher in the standards must be matched to that in the unknowns.

The overall effect of using the combination of a standard liquid scintillation spectrometer, a suitable program, and a computer is to give better data in a more usable form with little or no hand calculating.

REFERENCES

1. Peng, C. T., Atomlight 44 (March 1965).
2. Frank, R. B., Nuclear-Chicago Technical Data.
3. Blanchard, F. A., Intern. J. Appl. Radiation Isotopes 14:213 (1963).
4. Koehler, W. R., private communication.
5. Spratt, J. L., Intern. J. Appl. Radiation Isotopes 16:439 (1965).
6. Axelrod, L. R., Matthijssen, C., Goldzieher, J. W., and Pulliam, J. E., Acta Endocrinol. 49: Suppl. 99 (1965).
7. Matthijssen, C., Anal. Biochem. 15:382, (1966).
8. Baillie, L. A., Intern. J. Appl. Radiation Isotopes 8:1 (1960).
9. Bruno, G. A., and Christian, J. E., Anal. Chem. 33:1216 (1961).
10. Bush, E. T., Anal. Chem. 35:1024 (1963).
11. Rossel, A., Packard Instrument Co. Program 07661.
12. Beckman Bulletin 7100.

METHODOLOGY OF RADIOTRACER ENZYME ASSAYS

Donald J. Reed

Radiation Center and Department of Biochemistry and Biophysics
Oregon State University
Corvallis, Oregon

The inherent sensitivity and specificity of radiotracer enzyme assays has resulted in a rapid increase in their development and application. Also, as additional radiochemicals become commercially available, new assay procedures are being devised. Although radiochemicals have been utilized for enzyme assays for more than a decade, no review of the methodology of these assays has been published. Therefore, an attempt has been made in this article to illustrate the general methodology, versatility, advantages, and disadvantages of radiotracer enzyme assays. Specific comparisons of these assay methods with nonisotopic assay procedures are omitted due to space limitation.

DEFINITION OF ENZYME ACTIVITY

Enzyme activity is preferably described in terms of the International Unit [1]. One unit (U) of any enzyme is defined as that amount which will catalyze the transformation of 1 μmole of substrate per minute or, where more than one bond of each substrate molecule is attacked, 1 μequivalent of the group concerned per minute, under defined conditions. The temperature should be stated, and where practicable it should be 25°C. The other conditions, including pH, substrate concentration, where practicable should be optimal. The rate of an enzyme reaction is always understood as the initial rate. The initial rate or velocity should reflect an optimal steady-state concentration of the enzyme-substrate complex and its conversion to product at a maximal rate. Thus, reaction conditions should permit a maximal reaction rate proportional to the incubation time and enzyme concentration. Dixon and

Webb [2] present an excellent discussion of the main factors which influence enzyme activity, such as enzyme concentration, substrate concentration, pH, temperature, ionic medium, and the presence or absence of activators or inhibitors.

GENERAL METHODOLOGY

Radiotracer enzyme assays are based upon the enzymic conversion of a radioactive substrate to a radioactive product. The rate of the enzyme-catalyzed reaction is determined by radioassay of either the product, the unreacted substrate, or both after their separation from the assay reaction mixture. The specific activity of the substrate must be known since the total radioactivity in either the product or remaining substrate is used to calculate the quantity of product formed by enzymic activity.

TYPICAL RADIOTRACER ASSAY PROCEDURE

A typical radiotracer assay procedure consists of the following steps:

1. Preparation of a radioactive substrate of known specific activity.
2. Preparation of an enzyme source.
3. Incubation of the enzyme and radioactive substrate for a known period of time.
4. Incubation of one or more blanks with either boiled enzyme or without the enzyme.
5. Stopping of the reaction.
6. Separation of unreacted substrate and product.
7. Radioassay of an aliquot of the isolated product.
8. Calculation of the quantity of product formed as shown.

$$\frac{\text{Total dpm of} - \text{total dpm of the product}}{\text{isolated product} \quad \text{fraction of a boiled enzyme blank}}{\text{Specific activity of substrate (dpm/m}\mu\text{mole)}} = \begin{array}{c} \text{m}\mu\text{moles of product} \\ \text{formed per unit} \\ \text{time and enzyme} \end{array}$$

SEPARATION METHODS

Radiotracer enzyme assays depend on a quantitative separation of the substrate and the product formed. Often, only mμmoles or $\mu\mu$moles of product are formed by enzymic activity in many radiotracer assays. Thus, addition of unlabeled product and/or substrate to the assay mixture after the reaction is stopped often will aid in the quantitative separation of the product from remaining substrate.

Typical separation methods which have been used in these assays are listed in Table I. One can quickly conclude from the list of separation methods that while several of the procedures can be very rapid and convenient, certain of them are rather time-consuming. A more tedious radiotracer procedure for measuring a particular enzymic reaction is often used either because the method is extremely sensitive or because a more traditional assay method is not available or possible.

RADIOASSAY

A variety of counting methods are utilized in radiotracer enzyme assays. However, liquid scintillation counting is often the most convenient method of radioassay to establish the specific activity of the substrate and to measure the radioactivity in the product formed. Excellent articles have been published on liquid scintillation counting and on the appropriate fluor solutions for counting organic or water-soluble compounds [3-6].

ADVANTAGES OF RADIOTRACER ASSAYS

The merits of radiotracer assays are worthy of special mention. Shown in Table II are the main features of these assay procedures which should be stressed.

Table I.   Separation Methods for Radiotracer Assays

Volatilization	Thin-layer chromatography
Extraction	Paper chromatography
Extraction of a derivative	Electrophoresis
Precipitation	Column chromatography
Ion-exchange adsorption	Gel filtration
Charcoal adsorption	

Table II. Advantages of Radiotracer Assays

1. Extreme sensitivity

2. Specificity and its proof by isotope dilution analysis

3. Wide range of substrate and enzyme concentration

4. Minimal interferences during analysis

5. Measurement of low enzymic activity in crude homogenates

6. Competitive inhibition studies with substrate concentrations much less than the Km concentration

7. Effects of competitive inhibitors in crude homogenates minimized by dilution

The extreme sensitivity of radiotracer enzyme assays has not been fully realized or exploited in many applications of radiochemicals in these assays. Even so, a routine assay will often permit the measurement of formation of only a few mμmoles of product. The use of uniformly labeled substrates can extend the limits of sensitivity for the commonly used isotopes shown in Table III.

Half-life of the isotope and radiolysis are the main factors which set the practical limit on the sensitivity of radiotracer

Table III. Limit of Sensitivity of Radiotracer Assays

Radioisotope	Carrier-free specific activity (mc/matom)	Limit of Sensitivity* (moles of Product Formed)	
		Theoretical	Practical†
C^{14}	62	3×10^{-12}	3×10^{-12}
H^3	2.9×10^4	3×10^{-15}	1×10^{-14}
S^{35}	1.5×10^6	6×10^{-17}	1×10^{-12}
P^{32}	9.1×10^6	1×10^{-17}	1×10^{-12}
Cl^{36}	1.1	1×10^{-10}	5×10^{-10}

*Assuming that 220 dpm (0.1 m μc) is the lower limit of radioactivity in product to be radioassayed and that the substrate is specifically labeled, not uniformly labeled.
†Based upon currently available radiochemicals.

assays. These factors will be discussed in more detail in the next section.

The use of a radiochemical as a substrate often permits unequivocal proof of the specificity of the reaction being measured. Isotope dilution analysis and co-chromatography are two of the many techniques which can be brought to bear on this problem. Proof of specificity is an important factor when it is desirable to measure the enzyme activity of crude tissue homogenates.

Naturally occurring materials which are identical or similar to the product to be measured are often present in tissues at $\mu\mu$mole to $m\mu$mole concentrations. Frequently the variation in their concentration from tissue to tissue causes erratic blank values for many chemical methods of enzymic analysis. These substances may also influence the enzymic activity of the enzyme preparation. In addition, if the natural substance is identical to the radiochemical substrate, it can reduce the specific activity of the substrate. However, this effect can be minimized, as described in a following section. Thus, an important feature of radiotracer assays is that the analytical procedures of these assays do not place so great a limitation on conditions under which the reaction must proceed as do other methods for enzyme analyses. This advantage allows a wider range of substrate concentrations, changes in pH, ionic media, the presence and absence of activators and inhibitors which are chemically similar to the substrate, and so forth.

The sensitivity of radiotracer assays permits the assay of crude homogenates with low enzyme activity. Often these assays can be accomplished with microgram or milligram quantities of tissue. Thus, a survey of relative enzymic activity of even the hormone-secreting organs of a single biological specimen is often possible.

Effects of competitive inhibitors in crude homogenates may be greatly diminished by preparing dilute homogenates. Proportionality between the quantity of product formed and quantity of enzyme or time is often obtained only with crude homogenates prepared in dilute concentration.

Because of the sensitivity of radiotracer assays, competitive inhibition studies are often feasible, even when the substrate concentration is a thousandfold less than the Michaelis concentration (Km) of the enzyme.

Table   IV.   Disadvantages of Radiotracer Assays

---

1. Time-consuming separation of the unreacted substrate from the
    product prior to radioassay

2. Determination of the specific activity of the radiochemical
    substrate

3. Proof of purity of the radiochemical substrate

4. Reduction of the specific activity of the radiochemical
    substrate by endogeneous substrates

5. Not a continuous assay, for example, recording spectrophotometry

6. Possible isotope effects

DISADVANTAGES OF RADIOTRACER ASSAYS

The major disadvantages of radiotracer assays are summarized in Table IV.

The main undesirable feature of many radiotracer assays
is the time required to separate the unreacted substrate from
the product formed. However, since the labeled product as
well as the labeled substrate is often available, it is possible to
firmly establish the success of any particular separation
procedure. Addition of unlabeled substrate or product permits
one to adjust the chemical level to maximize the efficiency of
the separation method.

All too often the reputation of radiochemical companies is
the basis for accepting as fact the stated purity and specific
activity of a radiochemical. Indeed, many of the companies
are very reliable; however, the purity and specific activity
should be verified since the specific activity is the basis for
calculating the amount of product formed in the assay. If the
radiochemical is diluted with unlabeled carrier, the chemical
level of the radiochemical may be insignificant and the specific
activity can be based upon the total radioactivity of the radiochemical and the amount of unlabeled carrier added.

Radiochemicals become increasingly expensive when prepared with higher specific activities and greater degrees of
purity. At the higher specific activities, radiolysis or radiation self-decomposition becomes a more important factor which
contributes to the problems of purity. To minimize the cost of

the radiochemical and the effects of radiolysis, it becomes increasingly important to determine both the radiochemical and chemical levels of impurity which will not significantly affect the application of the particular radiochemical. Thus, defined purity of radiochemicals is of great importance to their application in radiotracer assays.

The cost of a radiotracer substrate can be a limiting factor in those radiotracer assays where maximum sensitivity is desired, and a high substrate concentration is necessary for optimal assay conditions. To reduce the cost, the assays are carried out in as small a volume as possible to minimize the total radiochemical needed. With the use of microsyringes, many assays are conducted with a total volume of 100 μliters or less.

Endogenous substrates in crude homogenates can decrease the specific activity of added radiochemical substrates to an unknown extent. Sephadex gel filtration can remove endogenous substrates without altering to a measurable degree the activity of some enzymes. Levine and Watts [7] have used this method successfully with L-histidine decarboxylase. Enzymes with reasonably high substrate Km values can be assayed with high substrate concentrations. This approach can be used to minimize the effect of unknown endogenous substrate concentrations. Also, some crude homogenates can be diluted sufficiently to essentially eliminate any effect due to endogenous compounds.

Radiotracer assays which utilize tritium-labeled substrates may show an isotope discrimination effect when the enzymically catalyzed reaction involves a tritium to carbon bond. Pomerantz [8] used tyrosine-C^{14} to demonstrate that tyrosine hydroxylase activity has little or no tritium rate effect when tyrosine-3,5-H^3 is employed as a substrate.

In the case of yeast aldolase, however, a large tritium isotope effect was seen in the condensation of tritiated dihydroxyacetone phosphate with glyceraldehyde-3-phosphate at pH greater than 7 [9]. Rose [10] has reviewed isotope effects as related to mechanisms of enzyme actions.

Finally, a word of caution should be given concerning isotopic exchange resulting in the conversion of isotope to the product without a net reaction. For example, the reaction catalyzed by creatine kinase has a rapid equilibrium, random mechanism [11]. Isotope exchange techniques for steady-state kinetic investigations have been described by Boyer and Silverstein [12].

Table V. Typical Examples of Radiotracer Assays

Enzyme	Substrate	Product	Separation method	Reference
Glycerol phosphate acyl transferase	P^{32} labeled glycerol phosphate + acyl CoA	P^{32} labeled glyceride monophosphate	precipitation and extraction	[13]
Propionyl CoA carboxylase	$K_2 C^{14}O_3$ + propionyl CoA	methylmalonyl CoA-C^{14}	volatilization of unreacted substrate	[14]
DNA polymerase	P^{32} labeled deoxynucleoside triphosphates	P^{32} labeled DNA	precipitation	[15]
UDP-glycose: β-1,3-glucan glucosyl transferase	UDP-D-glucose C^{14}	β1,3-glucan-C^{14}	extraction	[16]
Glutamic acid decarboxylase	L-glutamate-U-C^{14}	$C^{14}O_2$	volatilization	[17]
Homocysteine transmethylase	S adenosyl methionine-methyl-C^{14}	methionine-methyl-C^{14}	ion-exchange column	[18]

Table VI. Typical Examples of Radiotracer Assays

Enzyme	Substrate	Product	Separation method	Reference
Diamine oxidase	putrescine 1,4-C^{14}	Δ^1 pyrroline-C^{14}	extraction at pH 8.5	[19]
5-hydroxy-tryptophan decarboxylase	DL-5-hydroxy-tryptophan-C^{14}	5-hydroxy-tryptamine-C^{14}	extraction at pH 10	[20]
Steroid sulfokinase	S^{35} labeled 3-phosphoadenosine 5' phosphosulfate	S^{35} labeled steroid sulfate ester	precipita-tion	[21]
Sarcosine dehydrogenase	Sarcosine-methyl-C^{14}	C^{14} formalde-hyde	extraction of dimedon derivative	[22]
Succinate oxidase	Succinate-2,3-H^3	H_2^3O	ion exchange column	[23]
Acetylcholine esterase	Acetyl-1-C^{14} choline	Acetate-1-C^{14}	ion-exchange adsorption and extraction	[24]

TYPICAL EXAMPLES OF RADIOTRACER ASSAYS

In 1953 Kornberg and Pricer [13] described an assay of enzyme activity using a radioactive substrate. The assay was described as "an easy and extremely sensitive method for determining the extent of conversion of a water-soluble substance to a water-insoluble, fat-soluble substance: an isotopically labeled substrate of high specific activity is used in the incubation mixture." The assays listed in Tables V and VI illustrate examples of the versatility of radiotracer assays and some of the separation methods utilized.

SUMMARY

Radiotracer assay methods are well-established procedures for measuring enzyme activity. They are very versatile and extremely sensitive and generally have excellent reproducibility and accuracy. Their sensitivity often permits the use of microgram quantities of tissues to assay for the enzymic formation of $m\mu$mole to $\mu\mu$mole quantities of product. Radiotracer assay methods appear to represent a major advance in the methodology of enzyme assay procedures during the past decade.

REFERENCES

1. Report of the Commission on Enzymes of the International Union of Biochemists, Pergamon Press, Oxford, 1961.
2. Dixon, M., and Webb, E.C., Enzymes, Academic Press, New York, 1964.
3. Liquid Scintillation Counting, Bell, Jr., C.G., and Hayes, F.W., (editors), Pergamon Press, New York, 1958.
4. Advances in Tracer Methodology, Vol. 1, Rothchild, S. (editor), Plenum Press, New York, 1963.
5. Advances in Tracer Methodology, Vol. 3, Rothchild, S. (editor), Plenum Press, New York, 1966.
6. Wang, C.H., and Willis, D.L., Radiotracer Methodology in Biological Science, Prentice-Hall, Englewood Cliffs, New Jersey, 1965.
7. Levine, R.J., and Watts, D.E., Biochem. Pharmacol. 15:841 (1966).
8. Pomerantz, S.H., J. Biol. Chem. 241:161 (1966).
9. Rose, I.A., O'Connell, E.L., and Mehler, A.J., J. Biol. Chem. 240:1758 (1965).
10. Rose, I.A., Ann. Rev. Biochem. 35:23 (1966).
11. Morrison, J.F., and Cleland, W.W., J. Biol. Chem. 241:673 (1966).
12. Boyer, P.D., Arch. Biochem. Biophys. 82:387 (1959); Boyer, P.D., and Silverstein, E., Acta Chem. Scand. 17:Suppl. I, (1963).
13. Kornberg, A., and Pricer, Jr., W.E., J. Biol. Chem. 204:345 (1953); Methods of Enzymol. 1:673 (1955).

14. Flavin, M., Castro-Mendoza H., and Ochoa, S., J. Biol. Chem. 229:981 (1957).
15. Lehman, I.R., Bessman, M.J., Simms, E.S., and Kornberg, A., J. Biol. Chem. 233:163 (1958).
16. Feingold, D.S., Neufeld, E.F., and Hassid, W.Z., J. Biol. Chem. 233:783 (1958).
17. Albers, R.W., and Brady, R.O., J. Biol. Chem. 234:926 (1959).
18. Shapiro, S.K., and Yphantis, D.A., Biochim. Biophys. Acta 36:241 (1959).
19. Okuyama, T., and Kobayashi, Y., Arch. Biochem. Biophys. 95:242 (1961).
20. Snyder, S.H., and Axelrod, J., Biochem. Pharmacol. 13:805 (1964).
21. Wengle, B., Acta. Chem. Scand. 18:65 (1964).
22. Hoskins, D.D., and Bjur, R.A., J. Biol. Chem. 239:1856 (1964).
23. Goldsby, R.A., and Heytler, P.G., Biochim. Biophys. Acta 97:162 (1965).
24. Reed, D.J., Goto, K., and Wang, C.H., Anal. Biochem. 16:59 (1966).

BIBLIOGRAPHY OF RADIOTRACER ENZYME ASSAY*

I. Enzymes of Amine, Amino Acid, and Protein Metabolism

Enzyme	Radiochemical Substrate	Reference
Diamine oxidase	cadaverine 1,5-C^{14}	Okuyama, T., and Kobayashi, Y., Arch. Biochem. Biophys. 95:242 (1961).
Monoamine oxidase	tyramine-C^{14}	Otsuka, S., and Kobayashi, Y., Biochem. Pharmacol. 13:995 (1964).
Monoamine oxidase	tryptamine-2-C^{14}	Wurtman, R.J., and Axelrod, J., Biochem. Pharmacol. 12:1439 (1963).
Monoamine oxidase	5-hydroxytryptamine-C^{14}	McCaman, R.E., Federation Proc. 20:344 (1961).
	3-hydroxytyramine-C^{14} and tyramine-C^{14}	McCaman, R.E., McCaman, M.W., Hunt, J.M., and Smith, M.S., J. Neurochem. 12:15 (1965).
Dopamine-β-hydroxylase	dopamine-β-H^3	Goldstein, M., Prochoroff, N., and Sirlin, S., Experientia 21:592 (1965).

*This bibliography of radiotracer enzyme assays is not an exhaustive search of the literature for such assay methods. However, I do wish to gratefully acknowledge the assistance of Jacqueline Hucko, Russell Prough, Frank Taylor, Judith Wittkop, and Edith Yamasaki. The review of the literature was supported in part by the National Institutes of Health and the National Science Foundation.

Enzyme	Radiochemical Substrate	Reference
Ethanolamine deaminase	ethanolamine-1,2-C^{14}	Bradbeer, C., J. Biol. Chem. 240:4675 (1965).
Glutamic decarboxylase	L-glutamate-U-C^{14}	Albers, R.W., and Brady, R.O., J. Biol. Chem. 234:926 (1959).
3,4-dihydroxyphenyl-alanine decarboxylase	DL-dihydroxy-phenylalanine-2-C^{14}	Håkanson, R., and Owman, C.J., Neurochem. 12:417 (1965).
Acetyl aromatic amine transferase	acetate-1-C^{14}	Billiar, R.B., and Eik-Ness, K.B., Arch. Biochem. Biophys. 115:318 (1966).
Aspartate amino-transferase	2-oxoglutarate-5-C^{14}	Marino, G., Greco, A.M., Scardi, V., and Zito, R., Biochem. J. 99:589 (1966).
Glutamate trans-aminase	α-ketoglutarate-5-C^{14}	Waksman, A., Roberts, E., Arch. Biochem.Biophys. 109:522 (1965).
Sarcosine dehydrogenase	sarcosine-$C^{14}H_3$	Hoskins, D.D., and Bjur, R.A., J., Biol. Chem. 239:1856 (1964).
Tyrosinase	DL-tyrosine-2-C^{14} L-tyrosine-U-C^{14}	Chen, Y.M., and Chavin, W., Anal. Biochem. 13:234 (1965).
Tyrosinase	L-tyrosine-3,5-H^3	Pomerantz, S., Biochem. Biophys. Res. Commun. 16:188 (1964), J. Biol. Chem. 241:161 (1966).
Tyrosinase	tyrosine 3,5-H^3 tyrosine-U-C^{14}	Hall, P.F., and Okazaki, K., Biochem. 5:1202 (1966).
Tyrosine hydroxylase	L-tyrosine-3,5-H^3	Nagatsu, T., Levitt, M., and Uden-friend, S., Anal. Biochem. 9:122 (1964).
Glycine decarboxylase system	glycine-1-C^{14}	Klein, S., and Sagers, R.D., J.Biol. Chem. 241:197 (1966).
Histidine decarboxylase	histidine-1-C^{14}	Aures, D., and Clark, W.G., Anal. Biochem. 9:35 (1964).
Histidine decarboxylase	DL-histidine-1-C^{14}	Levine, R.J., and Watts, D.E., Biochem. Pharmacol. 15:841, (1966).
Tryptophan oxidative decarboxylase	L-tryptophan-3-C^{14}	Kosuge, T., Heskett, M.G., and Wilson, E.E., J. Biol. Chem. 241:3738 (1966).
Tryptophan synthetase	indole-C^{14}	Creighton, T.E., and Yanofsky, C., J. Biol. Chem. 241:980 (1966).

Enzyme	Radiochemical Substrate	Reference
5-hydroxytryptophan decarboxylase	DL-5-hydroxy-tryptophan-1-C^{14}	McCaman, R.E., McCaman, M.W., Hunt, J.M., and Smith, M.J., J. Neurochem. 12:15 (1965).
5-hydroxytryptophan decarboxylase	DL-5-hydroxy-tryptophan-C^{14}	Snyder, S.H., and Axelrod, J., Biochem. Pharmacol. 13:805 (1964).
Phenylalanine deaminase	DL-phenylalanine-3-C^{14}	Koukol, J., and Conn, E.E., J. Biol. Chem. 236:2692 (1961).
Iodotyrosine deiodinase	L-iodotyrosines-I^{131}	Falconer, I.R., Biochem. J. 100:197 (1966).
Glutamine synthetase	glutamate-C^{14}	Ravel, J.M., Humphreys, J.S., and Shive, W., Arch. Biochem. Biophys. 111:720 (1965).
Enzyme which converts tyrosine to phenol, pyruvate and NH_3	L-tyrosine-U-C^{14}	Brot, N., Smit, Z., and Weissbach, H., Arch. Biochem. Biophys. 112:1 (1965).
Carbamate kinase	P -ATP or Carbamyl-P^{32}	Marshall, M., and Cohen, P.P., J. Biol. Chem. 241:4197 (1966).
Carbamyl phosphate synthetase	NaH $C^{14} O_3$	Anderson, P.A., and Meister, A., Biochem. 4:2803 (1965)
Asparatic transcarbamylase	carbamyl phosphate-C^{14}	Issaly, A.S., and Reissig, J.L., Arch. Biochem. Biophys. 116:44 (1966).
Ornithine transcarbamylase	ureido-C^{14}-L-citrulline	Schimke, R.T., Berlin, C.M., Sweeney, E.W., and Carroll, W.R., J. Biol. Chem. 241:2228 (1966); Jones, M.E., Methods Enzymol. 5:903 (1957).
Ureidosuccinase	DL-carbamyl-C^{14} aspartate	Lieberman, I., and Kornberg, A., J. Biol. Chem. 212:909 (1955).
Argininosuccinate synthetase	L-ureido-C^{14} citrulline	Schimke, R.T., J. Biol. Chem. 239:136 (1964).
Argininosuccinase	guanido-C^{14} arginino succinic acid	Schimke, R.T., J. Biol. Chem. 239:136 (1964).
Arginase	L-guanido C^{14} arginine	Schimke, R.T., J. Biol. Chem. 239:136 (1964).
Urease	urea-C^{14}	Schimke, R.T., J. Biol. Chem. 239:136 (1964).
S-adenosylmethionine homocysteine methyltransferase	S-adenosyl L-methionine-$C^{14} H_3$	Shapiro, S.K., Yphantis, D.A., Biochim. Biophys. Acta 36:241 (1959).

Enzyme	Radiochemical Substrate	Reference
Methyltransferase	S-adenosyl methionine-$C^{14}H_3$	Axelrod, J., in Shapiro, S.K., and Schlenk, F., Transmethylation and Methionine Biosynthesis, University of Chicago Press, 1965.
S-adenosyl methionine synthase	L-methionine-$C^{14}H_3$	Mudd, S.H., Finkelstein, J.D., Irreverre, F., and Laster, L., J. Biol. Chem. 240:4382 (1965).
Homocysteine methyltransferase	S-methyl-L-methionine-$C^{14}H_3$	Abrahamson, L., and Shapiro, S.K., Arch. Biochem. Biophys. 109:376 (1965).
N^5-methyl-H_4-folate homocysteine methyltransferase	DL-N^5 $C^{14}H_3$ folate H_4	Weissbach, H., Peterkofsky, A., Redfield, B.G., and Dickerman, H., J. Biol. Chem. 238:3318 (1963).
Methyl-B_{12} - homocysteine methyltransferase	$C^{14}H_3$-B_{12}	Weissbach, H., Peterkofsky, A., Redfield, B.G., and Dickerman, H., J. Biol. Chem. 238:3318 (1963).
N^5-methyl-H_4-folate homocysteine methyltransferase	DL-N^5-methyl-C^{14}-H_4 folate	Brot, N., Taylor, R., and Weissbach, H., Arch. Biochem. Biophys. 114:256 (1966).
Methyl-B_{12}-homocysteine methyltransferase	B_{12}-$C^{14}H_3$	Brot, N., Taylor, R., and Weissbach, H., Arch. Biochem. Biophys. 114:256 (1966).
N^5-methyl tetrahydropteroyl-triglutamate homocysteine methyltransferase	N^5 methyl tetrahydropteroyl-triglutamate-C^{14}	Guest, J.R., and Woods, D.D., Biochem. J. 97:500 (1965).
O-methyltransferase	S-adenosyl methionine-$C^{14}H_3$	Tomita, K., Cha, C.M., and Lardy, H.A., J. Biol. Chem. 239:1202 (1964).
O-methyltranferase	S-adenosyl methionine-$C^{14}H_3$	Axelrod, J., and Weissbach, H., J. Biol. Chem. 236:211 (1961).
Catechol O-methylase	S-adenosyl methionine-$C^{14}H_3$	Daly, J., Inscoe, J.K., and Axelrod, J., J. Med. Chem. 8:153 (1965).
Histamine methyltransferase ·	S-adenosyl methionine-$C^{14}H_3$	Brown, D.D., Tomchick, R., and Axelrod, J., J. Biol. Chem. 234:2948 (1959).

Enzyme	Radiochemical Substrate	Reference
Phenethanolamine N-methyltransferase	S-adenosyl methionine-$C^{14}H_3$	Fuller, R.W., and Hunt, J.M., Anal. Biochem. 16:349 (1966).
Glutathione S-alkyltransferase	Iodomethane-C^{14}	Johnson, M.K., Biochem. J. 98:44 (1966).
Serine hydroxymethyltransferase	L-serine-U-C^{14}	Guest, J.R., and Woods, D.D., Biochem. J. 97:500 (1965).
Serine transhydroxymethylase	L-serine-3-C^{14}	Taylor, R.T., and Weissbach, H., Anal. Biochem. 13:80 (1965).
Cystathionine synthase	L-serine-3-C^{14}	Mudd, S.H., Finkelstein, J.D., Irreverre, F., and Laster, L., J. Biol. Chem. 240:4382 (1965).
Cystathionase	L-cystathionine-4-C^{14}	Mudd, S.H., Finkelstein, J.D., Irreverre, F., and Laster, L., J. Biol. Chem. 240:4382 (1965).
Aminoacyl s-RNA synthetase	amino acid-C^{14}	Makman, M.H., and Cantoni, G., Biochem. 4:1434 (1965).
Alanyl-s-RNA synthetase	pyrophosphate-P^{32}	Attwood, M.M., and Cocking, E.C., Biochem. J. 96:616 (1965).
Arginyl-ribonucleic acid synthetase	L-arginine-C^{14}	Allende, C.C., and Allende, J., J. Biol. Chem. 239:1102 (1964).
Leucyl-t-RNA synthetase	leucine-C^{14}	Yu, C.T., and Rappaport, H.P., Biochim. Biophys. Acta 123:134 (1966).
Prolyl-s-RNA synthetase	pyrophosphate-P^{32} (C^{14}) prolyl-s-RNA	Peterson, P.J., and Fowden, L. Biochem. J. 97:112 (1965).
Tyrosyl ribonucleic acid synthetase	pyrophosphate-P^{32}	Calendar, R., and Berg, P., Biochem. 5:1681 (1966).
D-ala-D-ala synthetase	D-alanine-C^{14}	Ito, E., and Strominger, J.L., J. Biol. Chem. 237:2689 (1962).
Phenylalanine activating enzyme	L-phenylalanine-C^{14}	Conway, T.W., Lansford, E.M., Jr., and Shive, W., Arch. Biochem. Biophys. 107:120 (1964).
Isoleucyl ribonucleic acid synthetase	PP_i-P^{32}	Berg, P., Bergman, F.H., Ofengand, E.J., and Dieckmann, M., J. Biol. Chem. 236:1726 (1961); Baldwin, A.N., and Berg, P., J. Biol. Chem. 241:831 (1966).

Enzyme	Radiochemical Substrate	Reference
D-glutamic acid adding enzyme	C^{14} D-glutamic acid CDP-GNAc lactyl-L-ala	Nathenson, S.G., Strominger, J.L., and Ito, E., J. Biol. Chem. 239:1773 (1964).
N-methylglutamate synthetase	methylamine-C^{14}	Shaw, W.V., Tsai, L., and Stadtman, E.R., J. Biol. Chem. 241:935 (1966).
Protein phosphokinase	ATP-P^{32}	Rabinowitz, M., and Lipmann, F., J. Biol. Chem. 235:1043 (1960).

II. Enzymes of Nucleic Acid Metabolism

Enzyme	Radiochemical Substrate	Reference
ATP-ADP exchange reaction	P^{32}-ADP, C^{14}-ADP	Wadkins, C.L., and Lehninger, A.L., J. Biol. Chem. 233:1589 (1958).
Pi-ATP exchange enzyme systems	Pi^{32}	Plaut, G.W.E., Methods Enzymol. 6:319 (1963).
ADP-ATP exchange enzyme	C^{14}-ADP	Rendi, R., Biochim. Biophys. Acta 118:621 (1966).
ATP-AMP phospho-transferase	C^{14}-AMP	Markland, F.S., and Wadkins, C.L., J. Biol. Chem. 241:4136 (1966).
ATP-Pi exchange enzyme	Pi^{32}	Skrede, S., Biochem. J., 98:702 (1966).
Adenosine triphosphatase	Pi^{32}	Crane, R.K., and Lipmann, F., J. Biol. Chem. 201:235 (1953).
Adenosine triphosphatase	ADP-C^{14}	Swanson, P.D., and Stahl, W.L., Biochem. J. 99:396 (1966).
Adenylic acid deaminase	adenosine 5' monophosphate-8-C^{14}	Hayashi, T.T., and Olmsted, P.S., Anal. Biochem. 10:354 (1965).

Enzyme	Radiochemical Substrate	Reference
Deoxycytidine monophosphate deaminase	dCMP-2-C^{14}	Shen, S.R., and Schmidt, R., Arch. Biochem. Biophys. 115:13 (1966).
Deoxycytidylate deaminase	dCMP-2-C^{14}	Maley, G.G., and Maley, F., J. Biol. Chem. 239:1168 (1964).
Endonuclease	P^{32}-labeled DNA	Lehman, I.R., Roussos, G.G., and Pratt, E.A., J. Biol. Chem. 237:819 (1962).
Endonuclease	P^{32}-labeled DNA	Linn, S., and Lehman, I.R., J. Biol. Chem. 240:1287 (1965).
DNA exonuclease	dAT-P^{32} copolymer	Richardson, C.C., and Lehman, I.R., J. Biol. Chem. 239:233 (1964).
DNA exonuclease II	P^{32}-labeled dAT copolymer	Lehman, I.R., and Richardson, C.C., J. Biol. Chem. 239:233 (1964).
DNA phosphatase-exonuclease	P^{32}-labeled native DNA	Richardson, C.C., Lehman, I.R., and Kornberg, A., J. Biol. Chem. 239:251 (1964).
Guanosine triphosphatase	GTP-γ-P^{32}	Conway, T.W., and Lipmann, F., Proc. Natl. Acad. Sci. U.S. 52:1462 (1964).
Adenylate kinase	AMP-C^{14} or ADP-C^{14}	Swanson, P.D., and Stahl, W.L., Biochem. J. 99:396 (1966).
Deoxynucleoside monophosphate kinase	P^{32}-deoxynucleotide	Lehman, I.R., Bessman, M.J., Simms, E.S., and Kornberg, A., J. Biol. Chem. 233:163 (1958).
Deoxynucleotide kinase	P^{32}-deoxynucleoside monophosphate	Sugino, Y., and Miyoshi, Y., J. Biol. Chem. 239:2360 (1964).
Deoxythymidine kinase	TdR-2-C^{14}	Bresnick, E., Thompson, U.B., and Lyman, K., Arch. Biochem. Biophys. 114:352 (1966).
Deoxythymidine kinase	C^{14} deoxythymidine	Okazaki, R., and Kornberg, A., J. Biol. Chem. 239:269 (1964).
5-phosphoribose pyrophosphokinase	C^{14} adenine	Kornberg, A., Lieberman, I., and Simms, E.S., J. Biol. Chem. 215:417 (1955).

Enzyme	Radiochemical Substrate	Reference
Thymidine kinase	thymidine-H^3	Behki, R.M., and Morgan, W.S., Arch. Biochem. Biophys. 107:427 (1964).
Thymidylate kinase	TMP-P^{32}	Grav, H.J., and Smellie, R.M.S., Biochem. J. 94:518 (1965).
Thymidine and thymidylate kinases	2-C^{14}-thymidine	Ives, D.H., Morse, Jr., P.A., and Potter, V.R., J. Biol. Chem. 238:1467 (1963).
Thymidylic kinase	H^3-TMP	Fausto, N., and Van Lancker, J.L., J. Biol. Chem. 240:1247 (1965).
Thymidine diphosphokinase	H^3-TDP	Fausto, N., and Van Lancker, J.L., J. Biol. Chem. 240:1247 (1965).
Thymidine kinase	deoxythymidine-2-C	Eker, P., J. Biol. Chem. 241:659 (1966).
Uridine kinase	uridine-2-C^{14}	Sköld, O., J. Biol. Chem. 235:3273 (1960).
DNA methylase	S-adenosyl methionine-$C^{14}H_3$	Oda, K., and Marmur, J., Biochem. 5:761 (1966).
RNA methylase	S-adenosyl-methionine-$C^{14}H_3$	Fleissner, E., and Borek, E., Biochem. 2:1093 (1963).
RNA methylase	S-adenosyl-methionine-$C^{14}H_3$	Hurwitz, J., Ander, M., Gold, M., and Smith, I., J. Biol. Chem. 240:1256 (1965); Gold, M., and Hurwitz, J., J. Biol. Chem. 239:3858 (1964).
Deoxyribonuclease	DNA-P^{32}	Lehman, I.R., Bessman, M.J., Simms, E.S., and Kornberg, A., J. Biol. Chem. 233:163 (1958).
Guanosine 5' triphosphatase	γP^{32}-GTP	Chan, M., and McCorquodale, D.J., J. Biol. Chem. 240:3116 (1965).
DNA phosphodiesterase	P^{32}-labeled DNA	Lehman, I.R., J. Biol. Chem. 235:1479 (1960)
DNA phosphatase	P^{32}-labeled 3' phosphoryl-terminated DNA	Richardson, C.C., and Kornberg, A., J. Biol. Chem. 239:242 (1964).
Deoxy CTP- and deoxy CDP splitting enzyme	d-C$P^{32}P^{32}$P	Kornberg, A., Zimmerman, S.B., Kornberg, S.R., and Josse, J., Proc. Natl. Acad. Sci. U.S. 45:772 (1959).

Enzyme	Radiochemical Substrate	Reference
RNase II	C^{14}-labeled poly A	Singer, M., and Tolbert, G., Bio-Chem. 4:1319 (1965).
Polynucleotide phosphorylase	$KH_2 P^{32} O_4$	Grunberg-Manago, M., Ortiz, P.J., and Ochoa, S., Biochim. Biophys. Acta 20:269 (1956).
Orotidine 5'-phosphate pyrophosphorylase	orotic acid $(4,7\text{-}C^{14})$	Lieberman, I., Kornberg, A., and Simms, E.S., J. Biol. Chem. 215:403 (1955).
Adenylate pyrophos-phorylase	adenine-8-C^{14}	Hori, M., and Henderson, J.F., J. Biol. Chem. 241:1406 (1966); Ellis, D.B., and Scholefield, P.G., Can. J. Biochem. Physiol. 40:343 (1962).
Guanosine 5'-PO_4 and inosine 5' PO_4 pyrophosphorylase	guanine-8-C^{14}	Kornberg, A., Lieberman, I., and Simms, E.S., J. Biol. Chem. 215:417 (1955).
Uridine 5'-phosphate pyrophosphorylase	uracil-2-C^{14}	Crawford, I., Kornberg, A., and Simms, E.S., J. Biol. Chem. 226:1093 (1957).
Purine nucleotide pyrophosphorylase	C^{14} purines	Davidson, J.D., Bradley, T.R., Roosa, R.T., and Law, L.W., J. Natl. Cancer Inst. 29:789 (1962).
Purine nucleotide pyrophosphorylase	C^{14} purine	Adye, J.C., and Gots, J.S., Bio-Chim. Biophys. Acta 118:344 (1966).
Cytidine diphosphate glucose pyrophosphorylase	glucose-C^{14}-1-PO_4	Kimato, K., and Suzuki, S., J. Biol. Chem. 241:1099 (1966).
DNA polymerase	P^{32}-deoxynucleoside triphosphates	Lehman, I.R., Bessman, M.J., Simms, E.S., and Kornberg, A., J. Biol. Chem. 233:163 (1958); Lehman, I.R., Methods Enzymol. 6:34 (1963).
DNA polymerase	C^{14} or P^{32}-labeled deoxynucleoside triphosphate	Okazaki, T., and Kornberg, A., J. Biol. Chem. 239:259 (1964).
DNA polymerase	P^{32}-TTP	Fausto, N., and Van Lancker, J.L., J. Biol. Chem. 240:1247 (1965).
DNA polymerase	P^{32} or C^{14}-labeled deoxynucleoside triphosphates	Orr, C.W.M., Herriott, S.T., and Bessman, M.J., J. Biol. Chem. 240:4652 (1965).

Enzyme	Radiochemical Substrate	Reference
RNA polymerase	ATP-8-C^{14}	Ochoa, S., Burma, D.P., Kroger, H., and Weill, J.P., Proc. Natl. Acad. Sci. U.S. 47:670 (1961).
RNA polymerase	C^{14} or P^{32} ribonucleoside triphosphates	Hurwitz, J., Methods Enzymol. 6:23 (1963).
RNA polymerase	P^{32} ribonucleoside triphosphates	Nakamoto, T., Fox, C.F., and Weiss, S.B., J. Biol. Chem. 239:167 (1964).
RNA polymerase	C^{14}-ATP	Stevens, A., and Henry, J., J. Biol. Chem. 239:196 (1964).
RNA polymerase	C^{14}-ATP	Weiss, S.B., Proc. Natl. Acad. Sci. U.S. 46:1020 (1960).
RNA polymerase	C^{14}-ATP or α-P^{32}-ATP	Mehrotra, B.D., and Khorana, H.G., J. Biol. Chem. 240:1750 (1965).
RNA polymerase	ATP-C^{14}	Mainwaring, W.I.P., and Williams, D.C., Biochem. J. 98:836 (1966).
RNA polymerase	ATP-C^{14}	Tata, J.R., and Widnell, C.C., Biochem. J. 98:604 (1966).
RNA polymerase	ATP-8-C^{14}	Nicholson, B.H., and Peacocke, A.R., Biochem. J. 100:50 (1966).
Cytidine triphosphate polymerase	H^3-CTP	Edmonds, M., J. Biol. Chem. 240:4621 (1965).
Polyriboadenylate polymerase	C^{14}-ATP or α-P^{32}-ATP	Smith, I., and August, J.T., J. Biol. Chem. 241:3525 (1966).
Ribonucleotide reductase	CDP-U-C^{14} or CDP-2-C^{14}	Hardy, J., and Beck, W.S., J. Biol. Chem. 240:3631 (1965).
Ribonucleotide reductase	CTP-2-C^{14}	Goulian, M., and Beck, W.S., J. Biol. Chem. 241:4233 (1966).
Adenylosuccinate synthetase	C^{14}-aspartate	Lieberman, I., J. Biol. Chem. 223:327 (1956).
Adenylosuccinate synthetase	L-aspartate-U-C^{14}	Hatch, M.D., Biochem. J. 98:198 (1966).
Adenylosuccinate synthetase	Adenylo-C^{14}-succinate	Hatch, M.D., Biochem. J. 98:198 (1966).

Enzyme	Radiochemical Substrate	Reference
Thymidylate synthetase	P^{32}-labeled dUMP	Friedkin, M., and Kornberg, A., in McElroy, W.D., and Glass, B. (editors), The Chemical Basis of Heredity, Johns Hopkins Press, Baltimore (1957).
Thymidylate synthetase	C^{14} formaldehyde	Flaks, J.G., and Cohen, S.S., J. Biol. Chem. 234:1501 (1959); Pizer, L.I., and Cohn, S.S., J. Biol. Chem. 237:1251 (1962).
Thymidylate synthetase	C^{14} formaldehyde	Blakley, R.L., and McDougall, B.M., J. Biol. Chem. 237:812 (1962).
Thymidylate synthetase	H^3-dUMP, H^3-TMP	Hartmann, K.U., and Heidelberger, C., J. Biol. Chem. 236:3006 (1961); Haggmark, A., Cancer Res. 22:568 (1962).
Thymidylate synthetase	P^{32}-dUMP, P^{32}-dTMP	Wahba, A.J., and Friedkin, M., J. Biol. Chem. 236:pc 11, (1961).
Thymidylate synthetase	deoxyuridine 5' mono-phosphate-5-H^3	Roberts, D., Biochem. 5:3546 (1966).
Xanthine oxidase	xanthine-8-C^{14}	Al-Khalidi, U.A.S., et al., Clin. Chim. Acta 11:726 (1965).
ADP-5-ribose phosphate adenyltransferase	Pi^{32}	Stern, A., and Avron, M., Biochim. Biophys. Acta 118:577 (1966).

III. Enzymes of Fatty Acid, Lipid, Terpene, and Steroid Metabolism

Enzyme	Radiochemical Substrate	Reference
Mevalonic kinase	mevalonate-C^{14}	Loomis, W.D., and Battaile, J. Biochim. Biophys. Acta 67:54 (1963).
Mevalonate kinase	DL-mevalonate-2-C^{14}	Williamson, I.P., and Kekwick, R.G.O., Biochem. J. 96:862 (1965).

Enzyme	Radiochemical Substrate	Reference
Squalene synthetase	farnesyl pyro-phosphate-4,8,12-C^{14}	Krishna, G., Whittock, Jr., H.W., Feldbruegge, D.H., and Porter, J.N., Arch. Biochem. Biophys. 114:200 (1966).
Propionyl-CoA CO_2 fixation	$K_2 C^{14} O_3$	Tietz, A., and Ochoa, S., J. Biol. Chem. 234:1394 (1965).
Prenol pyrophosphate pyrophospho-hydrolase	C^{14}-prenol-PP	Tsai, S.C., and Gaylor, J.L., J. Biol. Chem. 241:4043 (1966).
17α-hydroxylase	progesterone-4-C^{14}	Samuels, L.T., Short, J.G., and Huseby, R.A., Acta Endocrinol. 45:487 (1964).
17β-desmolase	17-hydroxypro-gesterone-4-C^{14}	Samuels, L.T., Short, J.G., and Huseby, R.A., Acta Endocrinol. 45:487 (1964).
17α-hydroxylase 17α-hydroxy pregene-C^{14}-C^{14}-lyase 17β-hydroxy steroid dehydro-genase	progesterone-C^{14}	Murota, S.I., Shikita, M., and Tamaolei, B.I., Biochim. Biophys. Acta 117:241 (1966).
9-α hydroxylase	androst-4-ene-3, 17-(4-C^{14}) dione	Chang, F.N., and Sih, C.J., Bio-chem. 3:1551 (1964).
3-β-hydroxy steroid dehydrogenase and Δ^5-3-keto steroid isomerase	Δ^5-3β-hydroxy steroids	Cheatum, S.G., and Warren, J.C., Biochim. Biophys. Acta 122:1 (1966).
Cholesterol esterase	(4-C^{14}) cholesteryl linoleate	Clarenberg, R., Steinberg, B., Asling, J.H., and Chaikoff, I.L., Biochem. 5:2433 (1966).
Cholesterol side-chain desmolase	cholesterol-26-C^{14}	Kimura, T., Satoh, P.S., and Tchen, T.T., Anal. Biochem. 16:355 (1966).
Progesterone oxygenase	progesterone-4-C^{14}	Rahim, M.A., and Sih, C.J., J. Biol. Chem. 241:3615 (1966).
Cholestenone 5α reductase	cholestenone-4-C^{14}	Shefer, S., Hauser, S., and Mosbach, E.H., J. Biol. Chem. 241:946 (1966).

Enzyme	Radiochemical Substrate	Reference
Steroid sulfatase	7α-H^3 dehydroisoandrosterone sulfate	Burstein, S., and Dorfman, R.I., J. Biol. Chem. 238:1656 (1963).
Lipopolysaccharide galactosyl transferase	UDP-galactose-C^{14}	Osborn, M.J., Rosen, S.M., Rothfield, L., and Horecker, B.L., Proc. Natl. Acad. Sci. 48:1831 (1962).
Polyglycerolteicholic acid glucosyltransferase	UDP-D-glucose-C^{14}	Glaser, L., and Burger, M.M., J. Biol. Chem. 239:3187 (1964).
Lipopolysaccharide N-acetylglucosaminyl transferase	UDP-N-acetyl-glucosamine	Osborn, M.J., and D'Ari, L., Biochem. Biophys. Res. Commun. 16:568 (1964).
Lipopolysaccharide glucosyl transferase	UDP-glucose-C^{14}	Rothfield, L., Osborn, M.J., and Horecker, B.L., J. Biol. Chem. 239:2788 (1964).
Acceptor phosphoglyceroltransferase	CDP-glycerol-C^{14}	Burger, M.M., and Glaser, L., J. Biol. Chem. 239:3168 (1964).
Glucosyl and galactosyl ceramide glycosidase	glucose-C^{14}-cerebroside galactose-1 C^{14}-cerebroside	Brady, R.O., Gah, A.E., Kanfer, J.N., and Bradley, R.M., J. Biol. Chem. 240:3766 (1965).
Purified phosphatidylethanolamine-N-methyl transferase	S-adenosyl methionine-methyl-C^{14}	Kaneshiro, T., and Law, J.N., J. Biol. Chem. 239:1705 (1964).
Phosphatidyl methyltransferase	S-adenosyl methionine methyl C^{14}	Rehbinder, D., and Greenburg, D.M., Arch. Biochem. Biophys. 109:110 (1965).
CDP glycerol and CDP ribitol pyrophosphorylases	P^{32}-sodium pyrophosphate	Crane, R.K., and Lipmann, F., J. Biol. Chem. 201:235 (1953).
Palmitoyl-CoA-L-glycerol 1-PO_4 palmitoyltransferase	glycerol phosphate glycerol 1-PO_4-P^{32}	Kuhn, N.J., and Lynen, F., Biochem. J. 94:240 (1965).
L-serine-CMP phosphatidyl transferase	DL-serine-3-C^{14}	Kanfer, J., and Kennedy, E.P., J. Biol. Chem. 239:1720 (1964).
Phosphatidyl serine decarboxylase	phosphatidyl L-serine-1-C^{14}	Kanfer, J., and Kennedy, E.P., J. Biol. Chem. 239:1720 (1964).
Cytidine diphosphate diglyceride synthetase	cytidine-H^3 triphosphate	Carter, J.R., and Kennedy, E.P., J. Lipid Res. 7:678 (1966).

Enzyme	Radiochemical Substrate	Reference
Phosphocholine-glyceride transferase	C^{14}-cytidine diphosphocholine	Weiss, S.B., Smith, S.W., and Kennedy, E.P., J. Biol. Chem. 231:53 (1958); McCaman, R.E., and Cook, K., J. Biol. Chem. 241:3390 (1966).
Phosphorylcholine cytidyl transferase	phosphorylcholine-methyl-C^{14}	Borkenhagen, L.F., and Kennedy, E.P., J. Biol. Chem. 227:951 (1957); Fiscus, W.G., and Schneider, W.C., J. Biol. Chem. 241:3324 (1966).
Choline acetyl-transferase	acetyl-1-C^{14} CoA	Schubert, J., Acta Chem. Scand. 17, Supp. I:233 (1963).
Choline acetyl-transferase	acetyl-1-C^{14} CoA or acetyl-1-C^{14} choline	McCaman, R.E., and Hunt, J.M., J. Neurochem. 12:253 (1965).
Choline acetyl-transferase	acetyl-1-C^{14}-CoA or acetyl-1-C^{14}-choline	Alpert, A., Kisliuk, R.L., and Shuster, L., Biochem. Pharmacol. 15:465 (1966).
Choline acetyl-transferase	acetate-1-C^{14}	Fonnum, F., Biochem. J. 100:479 (1966).
Acetyl transacylase	acetyl-1-C^{14}-CoA	Williamson, I.P., and Wakil, S.J., J. Biol. Chem. 241:2326 (1966).
Malonyl transacylase	malonyl-1, 3-C^{14}-CoA	Williamson, I.P., and Wakil, S.J., J. Biol. Chem. 241:2326 (1966).
Carnitine acetyl-transferase	acetyl-1-C^{14}-CoA	McCaman, R.E., McCaman, M.W., and Stafford, M.L., J. Biol. Chem. 241:930 (1960).
Carnitine-CoA-oleyltransferase	carnitine-C^{14}	Allmann, D.W., Galzigna, L., Mc-Caman, R.E., and Green, D.E., Arch. Biochem. Biophys. 117:413 (1966).
Lipoamidase	lipoic acid-S^{35}	Suzuki, K., and Reed, L.J., J. Biol. Chem. 238:4021 (1963).
Phosvitin kinase	ATP-P^{32}	Rodnight, R., and Lavin, B.E., Biochem. J. 93:84 (1964).
Lipase	C^{14}-triolein (carboxyl labeled)	Chino, H., and Gilbert, L.I., Anal. Biochem. 10:395 (1965).
Phospholipase C	C^{14}-lecithin	Graf, E., and Stein, Y., Biochim. Biophys. Acta 116:166 (1966).

Enzyme	Radiochemical Substrate	Reference
Acetyl cholinesterase	acetyl-1-C^{14} choline	Reed, D.J., Goto, K., and Wang, C.H., Anal. Biochem. 16:59 (1966).
Acetyl cholinesterase	acetyl-1 C^{14} choline	Winteringham, F.P.W., and Disney, R.W., Biochem. J. 91:506 (1964).
2-keto fatty acid decarboxylase	2-hydroxystearic acid-1-C^{14}	Davies, W.E., Hajra, A.K., Parmar, S.S., Radin, M.S., and Mead, J.F., J. Lipid Res. 7:270 (1966).
α-hydroxy acid decarboxylase	Stearic acid-1-C^{14}	Levis, G.M., and Mead, J.F., J. Biol. Chem. 239:77 (1964).

IV. Enzymes of Carbohydrate and Citric Acid Cycle Metabolism

Enzyme	Radiochemical Substrate	Reference
Carboxydismutase	$NaHC^{14}O_3$	Rabin, B.R., and Trowe, P.W., Proc. Natl. Acad. Sci. U.S., 51:497 (1964).
Ribulose diphosphate carboxylase	$KHC^{14}O_3$	Paulsen, J.M., and Lane, M.D., Biochem. 5:2350 (1966).
Pyruvate carboxylase	$NaHC^{14}O_3$	Nelson, P., Yarnell, G., and Wagle, S.R., Arch. Biochem. Biophys. 114:543 (1966).
α-keto glutarate carboxylase	$NaHC^{14}O_3$	Nelson, P., Yarnell, G., and Wagle, S.R., Arch. Biochem. Biophys. 114:543 (1966).
Pyruvate carboxylase	$KHC^{14}O_3$	Ruiz-Amil, M., deTorrontegui, G., Palacián, E., Catalina, L., and Losada, M., J. Biol. Chem. 240:3485 (1965).
Succinic oxidase	succinate-2, 3-H^3	Goldsby, R.A., and Heytler, P.G., Biochim. Biophys. Acta 97:162 (1965).
Phosphoenol pyruvate carboxy-kinase	$NaHC^{14}O_3$	Nelson, P., Yarnell, G., and Wagle, S.R., Arch. Biochem. Biophys. 114:543 (1966).

Enzyme	Radiochemical Substrate	Reference
Malate synthetase and other glyoxylate condensing enzymes	sodium glyoxylate-1-C^{14}	Wegener, W.S., Reeves, H.C., and Ajl, S.J., Anal. Biochem. 11:111 (1965).
Succinyl coenzyme A synthetase	C^{14}-succinyl phosphate	Nishimura, J.S., and Meister, A. Biochem. 4:1457 (1965).
UDP-D-galacturonic acid 4-epimerase	UDP-glucuronic acid-C^{14}	Feingold, D.S., Neufeld, E.F., and Hassid, W., J. Biol. Chem. 235:910 (1960).
N-acyl-D-glucosamine 6-phosphate 2-epimerase	C^{14}-N-acetyl glucosamine-6-P or C^{14}-N-acetyl mannosamine-6-P	Ghosh, S., and Roseman, S., J. Biol. Chem. 240:1531 (1965).
N-acylglucosamine 2-epimerase	C^{14}-labeled N-acetyl-glucosamine or C^{14}-labeled N-acetyl-mannosamine	Ghosh, S., and Roseman, S., Methods Enzymol. 8:191 (1966).
CDP-paratose 2-epimerase	CDP-paratose-C^{14}	Matsahashi, S., and Strominger, J.L., Biochem. Biophys. Res. Commun. 20:169 (1965).
a-1, 4 glucan hydrolases	malto-oligosac-charides-1-C^{14}	Pazur, J.H., and Okada, S., J. Biol. Chem. 241:4146 (1966).
GDP D-glucose glucohydrolase	GDP-C^{14}	Sonnino, S., Carminatti, H., and Cabib, E., Arch. Biochem. Biophys. 116:26 (1966).
Amino sugar kinases	C^{14}-amino sugars	Bates, C.J., and Pasternak, C.A., Biochem. J. 96:147 (1965); Buttin, G., Jacob, F., and Monad, J. Compt. Rend. Acad. Sci. Paris 250:247 (1960).
N-acyl-D-mannosamine kinase	N-acetyl or N-glycolyl-D-man-nosamine-C^{14}	Kundig, W., and Roseman, S., Methods of Enzymol. 8:195 (1966).
Fructokinase	fructose-C^{14}	Domagk, G.F., and Horecker, B.L., Arch. Biochem. Biophys. 109:342 (1965).
Galactokinase	galactose-C^{14}	Wilson, D.B., and Hogness, D.S., Methods Enzymol. 8,230 (1966).

Enzyme	Radiochemical Substrate	Reference
Galactokinase	galactose-C^{14}	Sherman, J.R., Anal. Biochem. 5:548 (1963); Sherman, J.R., and Adler, J., J. Biol. Chem. 238:873 (1963).
Glucose-6-P cyclase	C^{14}-glucose-6-P	Chen, I., Charalampous, F.C., J. Biol. Chem. 241:2194 (1966).
Phosphorylase Phosphatase	γP^{32}-ATP	Hurd, S.S., Novoa, W.B., Hickenbottom, J.P., and Fisher, E.H., Methods Enzymol. 8:546 (1966).
N-acylneuraminic (sialic) acid 9-phosphatase	C^{14}-N-acetylneuraminic acid	Jourdain, G.W., Swanson, A., Watson, D., and Roseman, S., Methods Enzymol. 8:205 (1966); J. Biol. Chem. 239:2714 (1964).
Sucrose phosphatase	sucrose phosphate (fructosyl-C^{14})	Hawker, J.S., and Hatch, M.D., Biochem. J. 99:102 (1966).
ADP-glucose pyrophosphorylase	sodium pyrophosphate-P^{32}	Preiss, J., Shen, L., Greenberg, E., and Gentner, N., Biochem. 5:1833 (1966).
ADP-glucose pyrophosphorylase	sodium pyrophosphate-P^{32}	Shen, L., and Preiss, J., Methods Enzymol. 8:262 (1966).
CDP-glucose pyrophosphorylase	glucose 1-phosphate-C^{14}	Mayer, R.M., and Ginsburg, V., J. Biol. Chem. 240:1900 (1965).
GDP-mannose pyrophosphorylase	C^{14}-D-glucose-1-P	Barber, G.A., and Hassid, W., Biochim. Biophys. Acta 86:397 (1964).
GDP-mannose pyrophosphorylase	pyrophosphate-P^{32}	Preiss, J., and Wood, E., J. Biol. Chem. 239:3119 (1964).
UDP-D-galactose D-glucose β-4-D galactosyltransferase	UDP-D-galactose-C^{14}	Watkins, W.M., and Hassid, W.Z., Biochem. Biophys. Res. Commun. 5:260 (1961).
UDP-D-galactose D-glucose β-4-D galactosyltransferase	UDP-D-galactose-1-C^{14}	Babao, H., and Hassid, W.Z., J. Biol. Chem. 241:2672 (1966).
Galactose 1-phosphate uridyltransferase	C^{14}-galactose-1-P	Bertoli, D., and Segal, S., J. Biol. Chem. 241:4023 (1966).
Gluosyltransferase	glucose-C^{14}	Torres, H.N., and Olavarría, J.M., J. Biol. Chem. 239:2427 (1964).

Enzyme	Radiochemical Substrate	Reference
β-1,3, glucan glucosyltransferase	UDP-D-glucose-C^{14}	Feingold, D.S., Neufeld, E.F., and Hassid, W.Z., J. Biol. Chem. 233:783 (1958).
D-xylodextrin xylosyl-transferase	UDP-D-xylose-C^{14}	Feingold, D.S., Neufeld, E.F., and Hassid, W.Z., J. Biol. Chem. 234:488 (1959).
Cellulose synthetase	GDP-glucose-C^{14}	Elbein, A.D., Barber, G.A., and Hassid, W.Z., J. Am. Chem. Soc. 86:309 (1964); Barber, G.A., Elbein, A.D., and Hassid, W.Z., J. Biol. Chem. 239:4056 (1964).
Sialyltransferase	C^{14}-CMP-N-acetyl-neuraminic acid	Jourdian, G.W., Carlson, D.M., and Roseman, S., Biochem. Biophys. Res. Commun. 10:352 (1963).
CMP-sialic acid synthetase (cytidine-5'-monophospho-sialic acid synthetase)	N-acetylneuraminic acid 1-C^{14} or N-glycolylneuraminic acid-1-C^{14}	Kean, E.L., and Roseman, S., Methods Enzymol. 8:208 (1966).
Colominic acid synthetase	CMP-N-acetyl neuraminic acid-C^{14}	Aminoff, D., Dodyk, F., and Roseman, S., J. Biol. Chem. 238:pc 1177 (1963).
ADP-glucose: α-1, 4-glucan-4-glucosyl transferase (trans-glucosylase)	ADP-glucose-C^{14} or deoxy-ADP glucose-C^{14}	Preiss, J., and Greenberg, D.E., Biochem. 4:2328 (1965).
Phytoglycogen synthetase	ADP-glucose-C^{14}	Frydman, R.B., and Cardini, C.E., Biochim. Biophys. Acta 96:294 (1965).
ADPG-starch trans-glucosylase (starch synthetase)	ADP-glucose-C^{14}	Murata, T., Sugiyama, T., and Akazawa, T., Arch. Biochem. Biophys. 107:92 (1964).
Uridine diphosphate glucose-glycogen glucosyltransferase	UDP-glucose-C^{14}	Steiner, D.F., Younger, L., and King, J., Biochem. 4:740 (1965).
α-1,4-glucan α4-glucosyltransferase (glycogen synthetase)	UDP-glucose-C^{14}	Robbins, P.W., Traut, R.R., and Lipmann, F., Proc. Natl. Acad. Sci. U.S. 45:6 (1959).
UDPG-glycogen trans-glucosylase	UDPG-C^{14}	Chandler, A.M., and Moore, O.R., Arch. Biochem. Biophys. 108:183 (1964).

Enzyme	Radiochemical Substrate	Reference
Sucrose synthetase	C^{14}-fructose	Slack, C.R., Phytochem. 5:397 (1966).
Transfructosylase	sucrose-(fructosyl-U-C^{14})	Edelman, J. and Dickerson, A.G., Biochem. J. 98:787 (1966).
GDP-L-fucose: lactose fucosyltransferase	GDP-L-fucose-C^{14}	Grollman, A.P., Hall, C., and Ginsburg, V., J. Biol. Chem. 240:975 (1965).
Polyribitol phosphate synthetase	CDP-ribitol-H^3 CDP-ribitol-P^{32} UDP-GlcNAc-C^{14}	Ishimoto, H., and Strominger, J.L., J. Biol. Chem. 241:639 (1966).
Uridine diphosphate D-glucose dehydrogenase	UDP-D glucose-C^{14}	Castanera, E., and Hassid, W.Z., Arch. Biochem. Biophys. 110:462 (1965).

V. Enzymes of Coenzyme and Mineral Metabolism

Enzyme	Radiochemical Substrate	Reference
N-methyl nicotin-amide oxidase	N-methyl (7-C^{14}) nicotinamide	Murashige, K., McDaniel, D., and Chaykin, S., Biochim. Biophys. Acta 118:556 (1966).
Nicotinic acid mononucleotide pyrophosphorylase	7-C^{14} nicotinic acid	Ogasawara, N., and Gholson, R.K., Biochim. Biophys. Acta 118:422 (1966).
Nicotinamide mononucleotide pyrophosphorylase	nicotinamide-C^{14}	Dietrich, L.S., Fuller, L., Yeoo, I.L., and Martinez, L., J. Biol. Chem. 241:188 (1966).
Nicotinic acid mononucleotide pyrophosphorylase	nicotinic acid-7C^{14}	Imsande, J., Preiss, J., and Handler, P., Methods Enzymol. 6:346 (1963).
Pyridone carboxamide forming enzyme	N'-methylnicotina-mide-7-C^{14}	Quinn, P., and Greengard, P., Arch. Biochem. Biophys. 115:146 (1966).
Nicotinic acid mononucleotidase	7-C^{14} nicotinic acid mononucleotide	Ogasawara, N., and Gholson, R.K., Biochim. Biophys. Acta 118:422 (1966).

Enzyme	Radiochemical Substrate	Reference
Nicotinamide deamidase	nicotinamide-7-C^{14}	Kirchner, J., Watson, J.G., and Chaykin, S., J. Biol. Chem. 241:953 (1966).
Quinolinate transphosphoribosylase	quinolinic acid-C^{14}	Nakamura, S., Ikeda, M., Tsuji, H., Nishizuka, Y., and Hayaishi, O., Biophys. Res. Commun. 13:285 (1963).
Quinolinated nicotinic acid mononucleotidase	quinolinic acid 2,3, 7,8,-C^{14}	Gholson, R.K., Veda, I., Ogasawara, N., and Henderson, L.M., J. Biol. Chem. 239:1208 (1964).
Folic acid reductase	folic acid-H^3	Rothenberg, S.P., Anal. Biochem. 13:530 (1965).
Dihydrofolate reductase	folate-H^3	Roberts, D., Biochem. 5:3549 (1966).
Trimethyl sulfonium-tetrahydrofolate methyltransferase	trimethyl sulfonium-C^{14}	Wagner, C., Lusty, Jr., S.M., Kung, H., and Rogers, N.L., J. Biol. Chem. 241:1923 (1966).
(+) biotin-apotrans-carboxylase synthetase	Na P^{32} Pi^{32}	Lane, M.D., Rominger, K.L., Young, D.L., and Lynen, F., J. Biol. Chem. 239:2865 (1964).
Sulfokinase	sodium sulfate^{35}	Wengle, B., Acta. Chem. Scand. 18:65 (1964).
Sulfotransferase	3'phosphoadenosine-5' phosphosulfate-S^{35}	Wengle, B., Acta. Chem. Scand. 18:65 (1964).
Monothiolphosphate hexose transferase	monothiolphosphate-P^{32}	Korman, E.F., Shaper, J.H., Cernichiari, O., and Smith, R.A., Arch. Biochem. Biophys. 109:284 (1965).
Glutathione-insulin transhydrogenase	I^{131}-insulin	Tomizowa, H.H., and Halsey, Y.D., J. Biol. Chem. 234:307 (1959).
Zinc-protoporphyrin chelatase	$Zn^{65}Cl_2$	Neuberger, A., and Tait, G.H., Biochem. J. 90:607 (1964).
Hydrogenase	H_2^3O	Anand, S.R., and Krasna, A.I., Biochem. 4:2747 (1965).
Metaphosphate synthetase	ATP-P^{32}	Kornberg, A., Kornberg, S.R., and Simms, E.S., Biochim. Biophys. Acta 20:215 (1956).

Enzyme	Radiochemical Substrate	Reference
ATP synthetase	P^{32}-polyphosphate	Kornberg, S.R., Biochim. Biophys. Acta 26:294 (1957).

THE USE OF LABELED DRUGS IN PSYCHO-PHARMACOLOGY RESEARCH

R. P. Maickel

Laboratory of Psychopharmacology
Departments of Pharmacology and Psychology
Indiana University
Bloomington, Indiana

A decade ago Gerard [1] published an article entitled "Drugs for the Soul; The Rise of Psychopharmacology." At that time, three out of ten prescriptions in the United States were being written for various tranquilizers, and the annual usage of a typical psychoactive drug, meprobamate, was in excess of 1 billion tablets. The use of such agents, not only in the treatment of severe mental illness, but also in ameliorating the mental stress of everyday life, has continued to mushroom. Along with the increased use of psychotherapeutic agents, there has been a concomitant growth of the research area known as "psychopharmacology."

Unfortunately, psychopharmacology is not a rigidly defined, limited discipline. Rather, it is a broad, interdisciplinary research area which includes psychology, pharmacology, physiology, biochemistry, organic chemistry, electronics, anatomy, psychiatry, clinical medicine, and many subdivisions of these areas. As a result, psychopharmacology research is extremely heterogeneous; the diversity may be easily recognized by the broad spectrum of scientific journals containing studies on various aspects of psychoactive drugs.

Despite this abundance of research, the widespread use of psychotherapeutic agents must be tempered by the understanding that we really know very little of how these drugs act. For example, Rees [2] recently stated: "Many of the important drugs that were responsible for the development of the modern era of psychopharmacology were discovered more by seren-

Supported by USPHS grant No. MH-06997.

dipity than by laboratory investigation." Although many clinically useful compounds have been discovered by accident, it is not a satisfactory feeling to realize that the therapy of mental disorders is largely dependent upon good fortune.

In order to satisfactorily examine the interactions of psychoactive drugs with biological organisms, one must consider the salient points to be attacked. As an experimental approach, these may be divided into four categories, each basically independent but with a considerable number of over-lapping and interlocking aspects:

1. The development of suitable techniques to measure animal behavior, looking at performance in areas such as ac-quisition, maintenance, retention, and extinction of learned behavior.
2. The examination of the effects of various psychoactive drugs on behavioral parameters, emphasizing dose—re-sponse relationships and drug—drug interactions.
3. The study of the physiological disposition of psychoactive drugs in the animal with particular emphasis on distribu-tion and localization within the central nervous system.
4. A consideration of the interaction of psychoactive drugs with biochemical and physiological systems in the animal, especially with neuronal mechanisms in the brain.

The present paper is concerned with the last two approaches, especially with regard to the advantages and disadvantages of using labeled drugs in such studies. The method of presentation will include experimental findings, a discussion of the dif-ficulties encountered, and the problems expected in future applications of the use of labeled drugs.

When physiological disposition studies are performed using labeled drugs, an absolute requirement is that the assay method used be highly specific. The final counting procedure should measure only the molecular species corresponding to unchanged drug. This requirement severely limits the use of autoadio-graphic techniques such as those described by Roth and Barlow [3] since such methods determine radioactivity but do not differ-entiate between parent compound and metabolic products. Methods of choice make use of suitable partition systems; liquid—liquid extraction, paper, thin-layer, or ion-exchange chromatography being most common. Thus, the biological

EXAMPLE 1

C^{14}:

   Background = 15 cpm

   Total count = 33 cpm

   C^{14} (net) = 18 cpm

$$\frac{background + sample}{background} = 2.2$$

Efficiency = 75%

18 CPM = 24 DPM

If the limit of sensitivity is to be 0.2 μg (2 × 10^{-7}g), then 24 dpm/0.2 μg = 120 dpm μg

Many drugs have a molecular weight of 200 to 300. If molecular weight = 250, then 120 dpm/μg = 30,000 dpm/μmole = 1.36 × 10^1 μc/mmole

Minimum sp. act. = 13.6 μc/mmole

To achieve a 5% counting error would require 2500 counts—or 76 min/sample.
To achieve a 2% counting error would require 14,000 counts—or 424 min/sample.

---

sample is treated in such a way that the parent compound is separated from various metabolic products prior to measurement of radioactivity.

Another related problem is that of sensitivity. Since most psychoactive drugs produce significant behavioral effects at relatively low dosages, assay procedures must be capable of accurately determining submicrogram quantities of drug. Furthermore, if one wishes to study distribution and localization of drugs in discrete areas of the central nervous system, assays must use small tissue samples. A drug present at a level of 1.0 μg/g of brain will yield only 0.01 ng from a sample of 10 mg of tissue. Example 1 shows the calculations for minimum specific activity demanded if one is using a compound labeled with C^{14} and counting with liquid scintillation techniques. From these values it is immediately apparent that C^{14} is not an ideal label for such studies. However, Example 2 shows that labeling with H^3 is much more satisfactory, since specific activities in excess of 1 c/mmole are usually attainable.

A typical problem may be described by our physiological disposition studies of the central nervous system stimulant, D-amphetamine. The chemical assay procedure for this drug, developed by Axelrod [4], lacks the necessary sensitivity for low dosage studies. However, with only slight modification,

EXAMPLE 2

H^3:

| Background = 8 cpm |
| Total count = 18 cpm |
| H^3 (net) = 10 cpm |

$$\frac{\text{background + sample}}{\text{background}} = 2.2$$

Efficiency = 18%
10 cpm = 56 dpm

If the limit of sensitivity is to be 0.05 μg (5×10^{-8}g), then 56 dpm/0.05 μg = 1120 dpm/μg.

Many drugs have a molecular weight of 200 to 300. If molecular weight = 250, then 1120 dpm/μg = 280,000 dpm/μmole = 127 μc/mmole.

Minimum sp. act. = 127 μc/mmole

To achieve a 5% counting error would require 2500 counts—or 139 min/sample.
To achieve a 2% counting error would require 14,000 counts—or 778 min/sample.

the procedure can be readily used for the assay of D-amphetamine-H^3. The structures of D-amphetamine and its major metabolic products are shown in Fig. 1. In order to remove the metabolic products, the following procedure was used: An aqueous phase consisting of tissue (homogenized in four volumes of 0.01 N HCl) or plasma was made alkaline (pH > 12) with 2 N NaOH and extracted with 5 to 6 vol. of benzene in glass-stoppered centrifuge tubes. After centrifugation, the aqueous phase was removed by aspiration and discarded. The benzene phase was then washed with 5 ml of 0.5 N NaOH. After centrifugation, a suitable aliquot of the benzene phase was transferred to a glass-stoppered centrifuge tube containing a small volume of

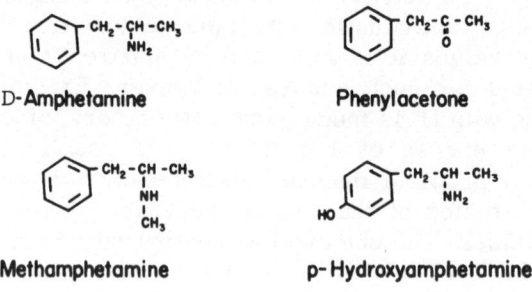

D-Amphetamine   Phenylacetone

Methamphetamine   p-Hydroxyamphetamine

Fig. 1

1 N HCl. After shaking, the tubes were centrifuged and an aliquot of the HCl phase was removed and assayed for H^3. All samples were assayed by liquid-scintillation procedures using polyethylene vials in a Packard Tricarb Model 4322 liquid scintillation counter. The counting system consisted of 0.5 ml of aqueous solution, plus 20 ml of a cocktail consisting of 14 g of BBOT and 280 g of naphthalene in a mixture of 2100 ml of toluene and 1400 ml of ethylene glycol monomethyl ether. This system gives 17 to 20% efficiency for H^3 with a background of 8 to 10 cpm. Using the extraction procedure described, the only possible metabolic product of D-amphetamine which would appear in the final HCl phase would be methamphetamine. This possibility was examined by thin-layer chromatography of acid phases from various tissues at different times after administration of D-amphetamine-H^3 to rats. Using silica gel G plates with a developing solvent of n-butanol:glacial acetic acid:water (4:1:5), only a single radioactive area was detected, corresponding in R_f to authentic D-amphetamine.

The data presented in Table I show results obtained after administration of D-amphetamine-H^3 (1 mg/kg, ip) to rats. The high tissue/plasma concentration ratios indicate extensive localization to tissue constituents. Further studies of the radioactivity contained in the various fractions of the assay procedure yields the data shown in Table II. From these data it is evident that a considerable portion of the radioactivity present is some tissues may consist of metabolic products of the drug. This confirms the lack of specificity of autoradiographic procedures to study drug disposition—such procedures

Table I. Physiological Disposition of D-Amphetamine-H^3
(generally labeled)*

| Time (hr) | D-Amphetamine-H^3 (ng/g) | | | | | | |
	Plasma	Brain	Fat	Heart	Kidney	Liver	Lung
0.5	378	2890	336	1237	9202	1809	6794
1.0	204	1488	181	633	5347	955	3934
2.0	71	640	68	348	2401	496	2377
4.0	23	190	18	99	685	138	803
8.0	7	51	9	27	220	51	231

*Each value is the mean of three rats given 1.0 mg/kg of D-amphetamine-H^3 (60 μc) by intraperitoneal injection at time zero.

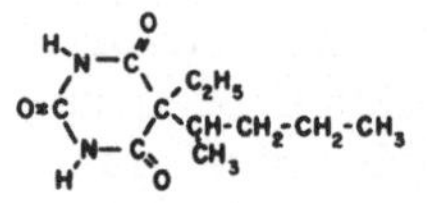

Pentobarbital

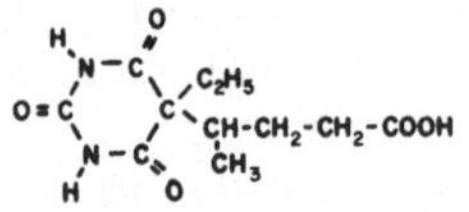

Pentobarbital carboxylic acid

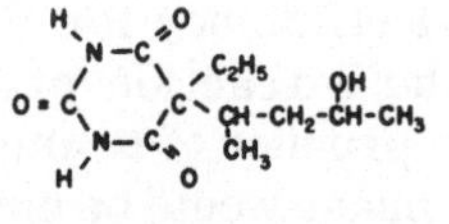

Pentobarbital alcohol

Fig. 2

measure the amount of radioactivity present in the tissues without regard for the chemical structures containing the radioactive isotope.

We have also studied the physiological disposition of C^{14}-pentobarbital. The primary metabolic products of this drug are shown in Fig. 2. The extraction procedure used for pentobarbital was a modification of that reported by Brodie et al. [5]. An aqueous phase consisting of 2 ml of homogenate (tissue + 3 vol 0.01 N HCl) or plasma, 1 g of NaCl, and 1.0 ml of pH 5.5 phosphate buffer was extracted with 10 vol of n-

Table II.   Radioactivity Present in Rat Tissues 90 Min After Administration of D-Amphetamine-H^3

Fraction	Percent of total H^3 in initial sample				
	Authentic D-amphetamine	-H^3 Plasma	Brain	Kidney	Liver
Aqueous residue (I)*	1.1	63.9	13.8	27.9	92.3
Alkali wash (II)	0.2	3.9	0.6	1.1	2.2
Benzene residue (III)	0.3	1.5	3.8	0.9	0.4
1 N HCl (IV)	98.4	30.8	81.6	70.2	5.4

*Fractions I and II should contain all of the p-hydroxyamphetamine; fractions I, II, and III should contain all of the phenylacetone; fraction IV may contain methamphetamine; conjugates of p-hydroxyamphetamine would remain in fraction I.

heptane containing 1.5% isoamyl alcohol. After centrifugation, the aqueous phase was removed by aspiration and discarded. The heptane was washed with 3.0 ml of pH 5.5 phosphate buffer, and, after centrifugation, an aliquot of the heptane phase was transferred to a clean shaking tube containing 20 ml of pH 11.0 phosphate buffer. After shaking and centrifugation, an aliquot of the aqueous phase was removed for determination of C^{14} using the liquid scintillation system described above. With 0.5 ml of aqueous phase and 20 ml of cocktail in polyethylene vials, background was 14-16 CPM with an efficiency or 65%.

The data in Table III show results obtained after administration of C^{14}-pentobarbital (9 mg/kg, ip) to rats. In contrast to D-amphetamine, pentobarbital is localized only in fat, kidney, and liver.

These typical results may be used to describe our approach to studying drug action on behavioral mechanisms. The plots of brain level against time for various doses of a drug can be compared with similar plots of behavioral effects against time. Thus, one can obtain a fairly accurate estimate of the level of drug in the brain that will procedure a significant decrement of behavioral performance. Table IV compares brain level of D-amphetamine as a function of time and dose with effects on a variety of behavioral tests. Several interesting observations may be made from this comparison. For example, systems in which the animal is tested continuously over a period of time (such as discriminated approach-avoidance and Sidman avoidance) appear to be more sensitive to drug effect than those test systems in which the animal is given only a single test

Table III.   Physiological Disposition of Pentobarbital-2-C^{14} *

	Pentobarbital-2-C^{14} (ng/g)						
Time	Plasma	Brain	Fat	Heart	Kidney	Liver	Lung
0.5	1054	897	2200	1105	2315	2257	1037
1.0	741	729	1489	737	1644	1470	606
2.0	399	377	858	378	1097	644	308
4.0	118	59	148	76	354	155	63
8.0	37	15	25	13	76	29	15

*Each value is the mean of 3 rats given 9.0 mg/kg of pentobarbital-2-C^{14} (10 μ c) by intraperitoneal injection at time zero.

Table IV.   Comparison of Brain Level of D-Amphetamine with
Behavior

Dose (mg/kg ip)	Brain level*	Behavior†	Brain level	Behavior†	Brain level	Behavior†	Brain level	Behavior†
0.25	812		412		127		39	
0.5	1496		816		397		84	
1.0	2890	V	1488	V	640	V	190	
2.0	5864	III, IV	2976	III, IV, V	1153	V	391	V
4.0	11293	III, IV	5732	III, IV	2372	III, IV	723	V
8.0	20697	I, II	10983	I, II	4710	I, II	1398	

*Brain level of D-amphetamine in $\mu g/g$.
†Significant deviation from placebo behavior in various tests as follows:
I.   Straightaway escape response [6] (not done at 0.25 mg/kg).
II. Hebb-Williams maze [7] (not done at 0.25 mg/kg).
III. Approach-avoidance system [8] (+ trials) (not done at 8 mg/kg).
IV. Approach-avoidance system [8] (- trials) (not done at 8 mg/kg).
V.   Continous (Sidman) avoidance [9] (not done at 8 mg/kg).

(such as the straightaway escape or Hebb-Williams maze). In
general, this type of comparison of brain levels of a drug with
its ability to impair behavioral performance would appear to
be an extremely valuable approach in elucidating the mechanism
of action of psychoactive agents.

However, the problem is not that simple. In many cases a
suitable labeled compound is not readily available. An excellent

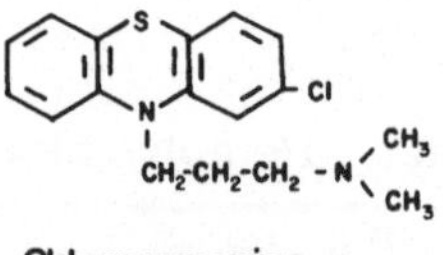

Chlorpromazine

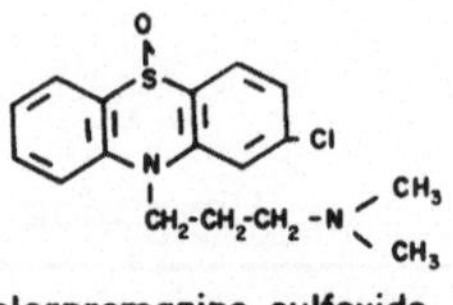

Chlorpromazine sulfoxide

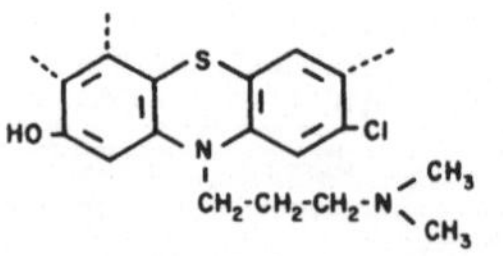

8-Hydroxychlorpromazine

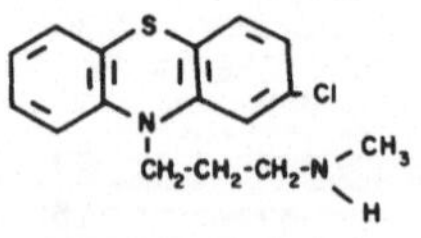

Desmethylchlorpromazine

Fig. 3

example of this type of difficulty may be seen in the tranquilizer chlorpromazine. The structure of this compound as well as those of some of its major metabolites are shown in Fig. 3. In the case of the hydroxylated metabolite, a number of different hydroxychlorpromazines have been isolated from the urine of animals receiving the drug. Presently two varieties of labeled chlorpromazine are commercially available:

1. S^{35}-ring sulfur: This compound is not satisfactory for three reasons. (1) The half-life of S^{35} is 87 days; thus, any long-term study requires the use of correction factors. (2) The location of the label is on an atom which is the site of a metabolic change. Although the atomic weight differential is only $+9.3\%$ (S^{35}/S^{32}), a possible isotope effect must be considered. In addition, the location of the tagged atom may cause chemical instability of the parent compound or the sulfoxide metabolite. (3) The specific activity is not sufficient for the sensitivity needed.

2. C^{14}-side-chain methyl: This compound is not satisfactory since one of the known metabolic pathways involves the loss of one (or perhaps both) of these methyl groups.

Thus, research on chlorpromazine awaits the availability of a label in a suitable position such as H^3 or C^{14} in the phenothiazine nucleus or in the n-propyl side chain. Similar problems exist for many other psychoactive drugs.

What is the future of labeled compounds in psychopharmacology research? The answer can only be that it is considered extremely bright. By using labeled compounds it should be possible to study the localization of such drugs in various areas of the central nervous system, perhaps even in specific functional areas concerned with specific behavioral phenomena. It is conceivable that such studies will lead to better understanding of the mechanism of drug action and eventually to knowledge of the effects of psychoactive drugs on these basic biological systems which control animal behavior. The research involved will require intensive collaboration from a number of diverse scientific disciplines, ranking from the organic chemist, who must prepare suitably labeled compounds, through the biomedical scientists, who must study the interaction of the compounds with the central nervous system, to the psychiatrist, who must apply the findings to the therapy of abnormal mental function.

REFERENCES

1. Gerard, R.W., Science 125:201 (1957).
2. Rees, L., in Marks, J., and Pare, C.M.B. (editors), The Scientific Basis of Drug Therapy in Psychiatry, Pergamon Press, Oxford, 1965.
3. Roth, L.J., and Barlow, C.F., in Roth, L.J. (editor), Isotopes in Experimental Pharmacology, University of Chicago Press, Chicago, 1965.
4. Axelrod, J., J. Pharm. Exp. Therap. 110:315 (1954).
5. Brodie, B.B., Burns, J.J., Mark, L.C., Lief, P.A., Bernstein, E. and Papper, E.M., J. Pharmacol. Exp. Therap. 109:26 (1953).
6. Russell, R.W., Basic Psychological Studies of the Effects of Incapacity Agents, Final Report, U.S. Army Chemical Center, July 31, 1962.
7. Rosvold, H.E., and Mirsky, A.F., Can. J. Psychol. 8:10 (1954).
8. Ray, O.S., Psychopharmacologia 4:326 (1963).
9. Heise, G.A., and Boff, E., Psychopharmacologia 3:264 (1962).

APPLICATION OF TRACERS TO QUANTITATIVE HISTOCHEMICAL AND CYTOCHEMICAL STUDIES

Richard E. McCaman

Institute of Psychiatric Research
Indiana University School of Medicine
Indianapolis, Indiana

INTRODUCTION

The biochemist, concerned with metabolic characteristics of specific cell types or of specific subcellular organelles, is all too frequently limited by the sensitivity of the methods for the assay of a particular chemical or enzyme constituent. This limitation in sensitivity is a function of many variables and would include (1) the inherent sensitivity of the physical technique used for measurements (pH, absorbance, fluorescence, radioactivity, and the like), (2) the concentration or activity of the component being studied, and (3) the size of tissue sample available for study (gram vs. microgram). The latter points may apply particularly to the problems encountered in an attempt to study the enzymatic and chemical constituents of cell types present in the nervous system [1-4].

Historically, quantitative microchemical studies such as those referred to above were dependent on absorbance (colorimetric) measurements in a spectrophotometer [5, 6]. More recently there has been a marked increase in the use of fluorescence measurements for the microquantitative estimation of a variety of biologic compounds [7]. The development of remarkably sensitive and relatively simple procedures for the fluorimetric measurement of pyridine nucleotides [8] has led to the development of other quantitative procedures for the determination of a wide variety of enzymes and substrates in very small tissue samples [9, 10].

Much of the author's research described in this article was supported by a grant (347) from the National Multiple Sclerosis Society.

While isotopically labeled compounds have a long history of use in qualitative and semiquantitative metabolic studies, they have been relatively little used in quantitative histochemical and cytochemical studies. It is the purpose of the present report to illustrate how isotopically labeled materials, in conjunction with scintillation-counting techniques, may be used to develop quantitative micromethods for measuring a variety of enzymes (and chemical constituents) in tissues. It will be shown by specific examples how isotopic techniques have a potential sensitivity, convenience, and reliability equal to or greater than that of colorimetric and fluorimetric techniques currently used in biochemical studies. Finally, certain unique advantages of isotopic techniques in quantitative histochemical and cytochemical studies will be presented.

COMPARATIVE SENSITIVITY OF PHYSICAL MEASUREMENTS

It is of interest, first, to compare the inherent sensitivity of microquantitative procedures, utilizing a scintillation spectrometer for the detection of labeled materials, to that for colorimetric or fluorimetric procedures currently used in the determination of a variety of enzymes and substrates of biochemical interest. One may arbitrarily define the "limit sensitivity" of any analytical procedure in terms of the amount of material that may be readily detected above the level of the blank (noise or background). For purposes of the present discussion, the "limit sensitivity" is defined as that amount (moles) of material that will give rise to a signal (counts, optical density, or fluorescence) which is twice that obtained from the "blank." Suppose, for example, that under a given set of conditions a reagent "blank" has 25 units of fluorescence. If 50 units of fluorescence are obtained from a sample containing 10^{-11} moles of some fluorogenic substance in the same reagent, the net fluorescence (50 - 25 = 25) is equal to that of the blank. By arbitrary definition, then, the "limit sensitivity" of such a substance is 10^{-11} moles.

The information used in the calculation of a "limit sensitivity" factor for a C^{14}-labeled compound assumed to have a specific activity of 1 mc/mmole is shown in Fig. 1. The value for "limit sensitivity" for a fluorimetric procedure is based

Physical Technique	"Limit Sensitivity"
Colorimetric:	0.5×10^{-9} moles (microcells)
Fluorimetric:	$10^{-10} - 10^{-11}$ moles (NADH, NAD)
Isotopic:	1.3×10^{-11} moles

Sample calculation for "limit sensitivity" of isotopic procedure:

$$10^{-3} \frac{mole}{mcurie} \times \frac{mc}{2.2 \times 10^9 \; dpm} \times \frac{dpm}{.7 \; cpm} = \frac{6.5 \times 10^{-13} \; moles}{cpm}$$

if "machine blank" = 20 cpm

then 1.3×10^{-11} moles = limit sensitivity

The values used for this calculation are: a) the reciprocal of the assumed specific activity, b) the constant relating dpm to millicuries, and c) an assumed counting efficiency of 70% for C^{14} in liquid scintillation counting procedures.

Fig. 1. Comparative sensitivity of various procedures.

on data obtained in the author's laboratory for the determination of pyridine nucleotides by the procedure of Lowry et al. [8]; for a spectrophotometric procedure, the determination of uric acid—molar absorbancy of approximately 1×10^7 mole^{-1} cm^2 at 292.5 mμ [6]—was used. It is apparent that the inherent sensitivity of the isotopic procedures is at least comparable to that of the fluorimetric procedures and greater than that of colorimetric procedures. However, it should be emphasized that the sensitivity of the isotopic procedures could be increased 10 to 100 times by using a radioactive substrate with a higher specific activity.

APPLICATION OF TRACERS TO ENZYME STUDIES

To achieve the stated level of sensitivity for isotopic procedures it is necessary to effect adequate separation of the labeled reaction product from the labeled substrate. Usually this must be done under conditions where the amount of labeled product represents 0.1 to 5% of the amount of labeled substrate. Further, the procedure involved should allow essentially quantitative recovery of the labeled product. For practical reasons this should also be a convenient and rapid procedure, so that one could analyze a moderate number of samples simultaneously.

Substrate	Product	Ref.
a) C¹⁴-Serotonin	5-OH-Indole acetaldehyde (acid)	[¹¹]
b) H³-Epinephrine (S-adenosylmethionine)	H³-Metanephrine	[¹²]
c) C¹⁴-Adenosylmethionine (Dihydroxybenzoate)	C¹⁴-3-Methoxy-4-hydroxy-benzoate	[¹³]
d) Cytidine diphospho-C¹⁴-choline (Diglyceride)	C¹⁴-Phosphatidyl choline	[¹⁵]
e) Cytidine diphospho-C¹⁴-ethanolamine (Diglyceride)	C¹⁴-Phosphatidylethanolamine	[¹⁶]
f) C¹⁴-L-α-Glycerolphosphate (Oleyl CoA)	C¹⁴-Phosphatidic acid	[¹⁷]
g) C¹⁴-Carnitine (Oleyl CoA)	C¹⁴-Oleylcarnitine	[¹⁸]

Fig. 2. Enzyme reactions which are measured isotopically after solvent extraction.

The enzyme reactions for which we have been able to develop microchemical methods using isotopic procedures are summarized in Figs. 2, 4, and 5. In addition, a few examples are included that have been developed in other laboratories that fit the classification of being both quantitative and microchemical procedures. The following examples were arbitrarily selected by the author to illustrate the usefulness of isotopic procedures and were not intended to cover all examples of the use of tracers for quantitative enzymology. These reactions have been subdivided to emphasize the physical technique involved in the eventual isolation of the radioactive product.

The methods for measuring the activities of the enzymes catalyzing the first group of reactions (Fig. 2) depend on a simple solvent extraction to effectively isolate the labeled product from the labeled substrate.

It might be profitable to consider in some detail the procedure for monoamine oxidase (Fig. 3) as a relevant example of the above category. In this example one has the option of using three different labeled substrates to study the enzyme reaction and, indeed, all three have been used to study this reaction in nervous tissues [¹¹]. The substrate, a primary amine, is oxidatively deaminated to yield the corresponding aldehyde as the primary product. However, in the presence of oxygen and the presence of an additional enzyme (aldehyde oxidase), some portion of the aldehyde is oxidized further to a secondary product (the corresponding acid). Because of the secondary or "side reaction," the study of this enzymatic step has been difficult in the past, particularly when either oxygen consumption or the appearance of aldehyde served as the measure of the reaction. However, with the isotopic procedures the secondary reaction presented no particular pro-

blem.   Both   the   carboxyl   group,   if   present,   as   well   as   the
aromatic hydroxyl would be undissociated at an acid pH.   Thus,
both the aldehyde and the acidic products are quantitatively
extracted into ethyl acetate.   This solvent is present as an
upper phase, and an aliquot of the upper organic layer may be
removed, washed rapidly with another small volume of HCl,
and placed in a scintillation vial for counting.   The method is
quite simple and has a coefficient of variation of 2 to 3%.   The
"limit sensitivity" for this protocol is 10^{-11} moles, using
labeled serotonin which has specific activity of 1 mc/mmole.
The reagent blank is essentially background since the labeled
serotonin and tyramine are not soluble in ethyl acetate at an
acid pH.   Various lots of dopamine from different suppliers
at times have given us problems because of trace contaminants
which are labeled and which extract from an acidified aqueous
solutions into ethyl acetate.

This procedure serves to illustrate another useful advantage
of the isotopic procedures. They may be used with considerable
ease to study the effects of various antimetabolites or enzyme
inhibitors.   The quantitative effect of adding any unlabeled
compound is directly reflected in the amount of labeled product
formed except, of course, where the added material in some way
interferes with the physical method of recovering the labeled
product.   It should be further emphasized that in a study of
competitive inhibitors of enzyme activity it is desirable to
conduct these tests at a low substrate concentration.   The great

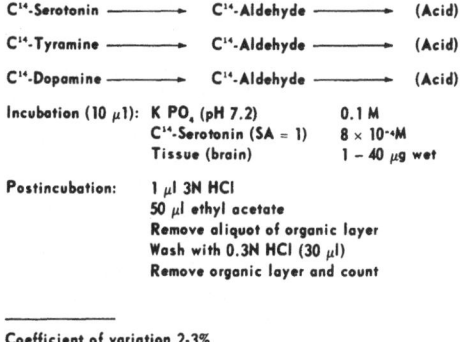

Fig. 3. Outline for the isotopic procedure for measuring monoamine oxidase activity.

sensitivity of the isotopic methods greatly facilitates such studies [11].

The procedure for measuring catechol-O-methyl transferase (Fig. 2b), developed by Axelrod et al. [12], utilizes H^3-epinephrine and S-adenosyl methionine as substrates. The product of the enzyme reaction, H^3-metanephrine, was extracted into a toluene-isoamyl alcohol solvent system. We experienced a number of difficulties trying to scale this procedure down to a more micro level. These problems were ultimately resolved by utilizing C^{14}-S-adenosyl-methionine as the labeled substrate and dihydroxybenzoic acid [13]. Axelrod and Tomchick [14] had previously shown that the latter was actually more active than epinephrine as a substrate for this enzyme. Again, the functional groups (COOH and phenolic hydroxyl) of the enzyme product, O-methyl-C^{14}-benzoic acid, are undissociated at an acid pH. Thus, the labeled product is readily extracted into an organic solvent such as ethyl acetate while C^{14}-S-adenosyl-methionine (substrate) is not extracted. The blank is very low and essentially quantitative recovery of the O-methyl derivative is obtained. As a result of a more efficient isolation of the labeled product and the somewhat greater enzyme activity with benzoic acid, a 1000 times increase in sensitivity was effected [13].

The remaining examples of this category (Fig. 2, d to g) have an important feature in common. Each of these reactions gives rise to a labeled lipid product which is quantitatively extracted into an organic solvent. Furthermore, in each case the labeled substrate is highly water soluble. Thus the methods for these reactions are characterized by very low reagent blanks [15—18].

Solvent extraction represents the simplest and most rapid physical technique for the isolation of a radioactive product. In the above examples it provides essentially quantitative recovery of the labeled products. While this is not an essential prerequisite, it should obviously allow a high recovery (> 80%) if maximum sensitivity is desired. It is also highly desirable that the solvent and the labeled substrate be selected to give little or no blank. An additional advantage of these techniques is that the solvent may usually be evaporated, thus minimizing any effect (quenching) on the scintillation counting.

The second category of reactions (Fig. 4) are measured by "microdiffusion" methods. The first three of these reactions

Substrate	Product	Ref.
a) C^{14}-Glutamic	$C^{14}O_2$	[19]
b) C^{14}-5-OH-Tryptophan	$C^{14}O_2$	[11]
c) C^{14}-6-Phosphogluconate (TPN)	$C^{14}O_2$	[20]
d) C^{14}-Acetate (choline, CoA, ATP)	$\underline{C^{14}\ Acetyl\ Choline}$	[21]

Fig. 4. Enzyme reactions which are measured isotopically by gaseous diffusion methods.

lead to the evolution of $C^{14}O_2$ and include enzyme-catalyzed reactions for glutamic decarboxylase [19], 5-hydroxytryptophan decarboxylase [11], and 6-P-gluconic dehydrogenase [20]. To the author's knowledge, the method for glutamic decarboxylase by Albers and Brady [19] was the first specific application of isotopes to quantitative microchemical studies. It should also be emphasized that these "gasometric" isotopic procedures are applicable to the study of any decarboxylation reactions, providing that the appropriate carboxyl-labeled substrates can be obtained or synthesized.

It may also be appropriate to include in this category a recently published procedure for the microdetermination of choline acetylase [21]. In the latter procedure the labeled substrate is C^{14}-acetate and the incubation mixture includes in addition to choline an acetyl CoA generating system (CoA, ATP, and enzyme). After the reaction is terminated by acidification, the sample is taken to dryness during which the C^{14}-acetic acid is volatilized. The evaporation may be carried out again with carrier acetic acid being added the second time. Thus, the nonvolatile labeled material is equivalent to the synthesized C^{14}-acetylcholine. This is an example of removing the substrate rather than isolating the product of the enzyme reaction.

A secondary chemical reaction is utilized to facilitate a measure of the enzyme catalyzed reactions shown in Fig. 5.

Substrate	Product (secondary reaction)	Ref.
a) C^{14}-Acetyl CoA (choline)	C^{14}-Acetylcholine (reinecke)	[22]
b) C^{14}-Acetyl CoA (carnitine)	$\underline{C^{14}\text{-Acetylcarnitine (periodide)}}$	[23]
c) $\underline{C^{14}\text{-Choline}}$ (ATP)	C^{14}-Choline phosphate (reinecke)	[15]
d) $\underline{C^{14}\text{-Ethanolamine}}$ (ATP)	C^{14}-Ethanolamine phosphate (DNFB)	
e) $\underline{C^{14}\text{-Adenosylmethionine}}$ (Norepinephrine)	C^{14}-Epinephrine (reinecke)	[24]

Fig. 5. Enzyme reactions which are measured isotopically by procedures involving a secondary chemical reaction.

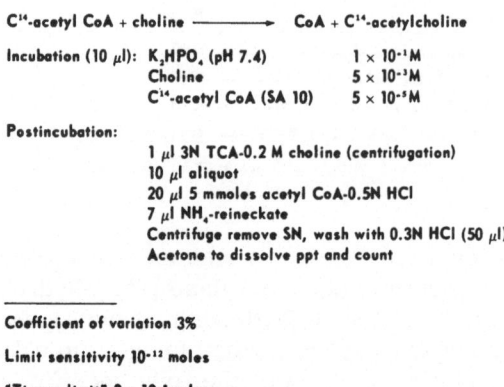

C^{14}-acetyl CoA + choline ⟶ CoA + C^{14}-acetylcholine

Incubation (10 μl): K₂HPO₄ (pH 7.4) 1×10^{-1} M
                      Choline 5×10^{-3} M
                      C^{14}-acetyl CoA (SA 10) 5×10^{-5} M

Postincubation:

       1 μl 3N TCA-0.2 M choline (centrifugation)
       10 μl aliquot
       20 μl 5 mmoles acetyl CoA-0.5N HCl
       7 μl NH₄-reineckate
       Centrifuge remove SN, wash with 0.3N HCl (50 μl)
       Acetone to dissolve ppt and count

Coefficient of variation 3%

Limit sensitivity 10^{-12} moles

"Tissue limit" 2×10^{-8} g dry

Fig. 6. Outline of procedure for the measurement of choline acetylase activity.

In the first two examples, the labeled products are converted to insoluble derivatives to facilitate their selective isolation from the respective labeled substrates. Conversely, the unmetabolized labeled substrate is selectively removed in the methods for the last three reactions.

The procedure for the quantitative microdetermination of choline acetyltransferase [22] is shown in some detail (Fig. 6). The labeled substrate is C^{14}-acetyl CoA and the product formed during the course of the enzyme reaction is C^{14}-acetylcholine. Once the enzymatic reaction is terminated, the radioactive product must be separated from the remaining radioactive substrate (a problem which is not unique to isotopic procedures). In this case, the C^{14}-acetylcholine is isolated as an insoluble reinecke salt by coprecipitation in the presence of an excess of choline (which is also a substrate in the reaction). To reduce the contamination of the radioactive product so isolated, unlabeled acetyl CoA (20 to 100 times excess) was added during the postincubation phase. The "reagent blank" thus obtained was generally only a few cpm above what we have referred to previously as the "machine blank." Since the substrate had a specific activity of 10 mc/mmole, our "limit sensitivity," both theoretical and practical, was approximately 10^{-12} moles of C^{14}-acetylcholine. The procedure is reasonably convenient and highly reproducible, giving a coefficient of variation of 3%. The protocol shown was used routinely for the analysis of choline acetyltransferase activity in samples of nervous tissue weighing 0.2 to 20 μg dry weight [22].

There are at least two side reactions which could affect the quantitative measure of this enzyme: (1) the hydrolysis of C^{14}-acetyl CoA (giving rise to C^{14}-acetate and CoA) and (2) the hydrolysis of C^{14}-acetylcholine. The enzyme catalyzing the latter reaction was completely inhibited by the addition of eserine to the incubation mixture. The first of these two reactions was of no consequence as long as the concentration of C^{14}-acetyl CoA was high enough to saturate the choline acetyltransferase during the incubation period.

This particular protocol may be used to illustrate another general point with regard to the isotopic procedures. The expense of a procedure is an important factor. C^{14}-acetyl CoA is one of our more expensive substrates. However, the quantity actually used per analysis (approximately 10^{-9} moles) working at the micro level allows several thousand analyses to be completed with 50μc of substrate.

Since many compounds containing a quaternary nitrogen form insoluble reinecke derivatives, a similar procedure was expected to be useful for the measurement of carnitine acetyl transferase, but the reineckate of acetylcarnitine is extremely water soluble. However, C^{14}-acetylcarnitine, in the presence of substrate quantities of unlabeled carnitine may be isolated quantitatively as an insoluble periodide [23]. There is very little reagent blank since acetyl CoA does not readily contaminate the periodide precipitate.

Thus far, methods have been described in which the procedural approach has involved the selective isolation of the enzymatic product free of the substrate. A procedure has been developed for the measurement of choline phosphokinase (Fig. 7) in which the separation is accomplished by elimination of the substrate [15]. In this procedure, C^{14}-choline and ATP are the substrates and the enzymatic product of interest is C^{14}-choline phosphate. The unmetabolized C^{14}-choline can be removed very conveniently as an insoluble reinecke, while C^{14}-choline phosphate remains completely soluble at an alkaline pH. As a postincubation procedure, sodium hydroxide is added, followed by the ammonium reinecke addition, and the C^{14}-choline is removed in the precipitate. A measured aliquot of the supernatant is taken and subjected to another addition of reinecke in the presence of additional carrier choline. The radioactivity in the resultant supernatant due to the C^{14}-choline is thus

C^{14}-choline + ATP ──────→ ADP + C^{14}-choline PO$_4$

Incubation (10 μl): Aminopropanol (pH 10.2) 1×10^{-1}M
                        ATP 1×10^{-2}M
                        C^{14}-choline (SA 0.75) 2×10^{-2}M

Postincubation:

                        20 μl 0.4N NaOH
                        10 μl NH$_4$ reineckate
                        centrifuge-save aliquot of SN
                        5 μl choline 1 $\times$ 10^{-1}M
                        10 μl NH$_4$ reineckate
                        centrifuge-aliquot SN to count

─────────────

Coefficient of variation 3%

Limit sensitivity 10^{-11} moles

Fig. 7. Outline of procedure for the measurement of choline phosphokinase activity.

reduced to zero (that is, the blank in this reaction is essentially that of the "machine blank," about 20 to 25 cpm). This is a highly reproducible procedure with a coefficient of variation of about 3%, and the limit sensitivity is 10^{-11} moles using C^{14}-choline with a specific activity of 0.75 mc/mmole. It is now possible to obtain labeled choline which has a specific activity of 20 to 30 mc/mmole.

The elimination of the substrate was also the most effective way of measuring two other enzymes, ethanolamine phosphokinase (unpublished procedure, R.E. McCaman) and N-methyltransferase [24]. In each case the labeled substrate was removed after the incubation, leaving behind the labeled product of the respective enzyme reactions.

APPLICATION OF TRACERS TO ENDOGENOUS SUBSTRATE MEASUREMENTS

Only a few examples have appeared involving the application of tracers for the assay of endogenous tissue constituents [25—27]. Nevertheless, the authors of these elegant techniques have provided a thorough theoretical and practical framework of considerations that should allow and encourage considerable extension of these studies in the development of additional methods. These procedures involve a double label assay technique utilizing auxillary enzyme systems. The assay of tissue levels of S-adenosylmethionine [25] is illustrated in Fig. 8. A portion of a deproteinized solution of tissue extract (containing an unknown amount of S-adenoslymethionine) is mixed with a known small amount (one tenth or less than the amount present in unknown) of S-adenoslymethionine-methyl-C^{14}. After the addition of a buffer, a known amount of N-acetyl-serotonin-H^3 is added along with partially purified hydroxy-indole-O-methyl transferase. During the course of incubation (37°), melatonin-methoxy-C^{14}-acetyl-H^3 is formed. After the reaction is terminated the doubly labeled product is selectively extracted into an organic solvent system and assays for C^{14} and H^3 are completed. A plot of the H^3/C^{14} ratio in the enzymatically generated melatonin against known quantities of unlabeled S-adenosylmethionine (standards) produces a linear function from which the unknown adenosylmethionine content of the tissues can be precisely calculated.

The procedure is quite sensitive ($< 1\mu g$ adenosylmethionine) and actually may be carried out with crude pineal gland homogenate (after dialysis to remove endogenous S-adenosyl-methionine) as the source of methylating enzyme. The specificity of the method is related to the selective utilization of S-adenosylmethionine for this methylation by hydroxyindole O-methyl transferase and the relative absence of tissue hydroxyindoles. With only a slight variation in the approach (destruction of endogenous tissue levels of adenosylmethionine by heating), the tissue levels of N-acetylserotonin (melatonin) may be determined using the hydroxyindole-O-methyl transferase (Baldessarini, unpublished observations).

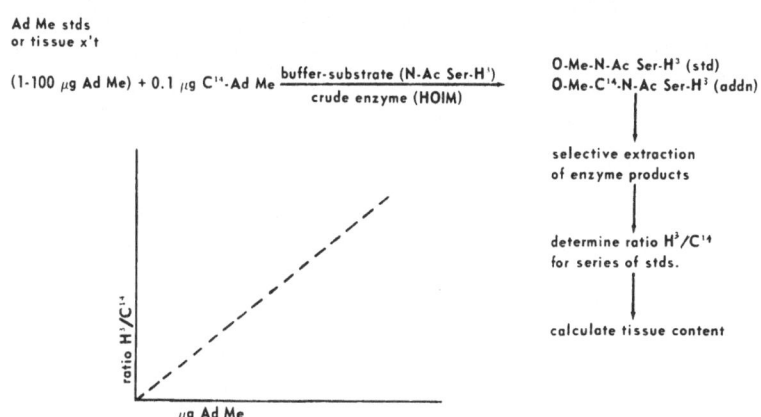

Fig. 8. Outline of procedure for isotopic determination of tissue levels of S-adenosylmethionine.

A very sensitive assay for tissue histamine also has been described, using a similar double label assay in conjunction with a partially purified histamine methyl transferase [27]. The C^{14}/H^3 ratio of the C^{14}-methylhistamine-H^3 was shown to be proportional to the histamine content of the tissues over the range of 0.002 to 1.00μg! The precision of the procedure is very good since the mean difference of duplicate determinations is only about 2%.

While the use of a single isotope in these procedures would be an obvious simplification, the above authors have observed no consistent relationship between the tissue concentration of a given constituent and the amount of the corresponding single-labeled enzymatic product [26].

These elegant procedures are not only very sensitive and specific but offer certain unique advantages because they utilize isotopic compounds. For example, the enzyme reaction does not have to be linear since one label (H^3) indicates the degree of dilution by the endogeneous tissue constituent and the other (C^{14}) actually monitors the degree of reaction. Furthermore, these procedures do not require the careful and laborious purification of tissues and reagents generally characteristic of other (that is, fluorimetric) procedures. (For further theoretical consideration of these procedures see Baldessarini and Kopin [26] and Snyder et al. [27].)

PROCEDURES FOR INCREASING THE SENSITIVITY OF ISOTOPIC PROCEDURES

Up to this point the author has been attempting to illustrate the utility of isotopes in the development of a variety of micro-quantitative methods. Let us turn to a brief consideration of ways in which the sensitivity may be further increased so that such procedures might be used, for example, in the study of the chemistry of a single cell. There are a number of possible ways to increase the sensitivity of isotopic methods, for example, the use of higher specific activities. The theoretical maximum, however, is related to multiples of that obtained for pure C^{14}, that is, 62 mc/mmole. But there are practical limitations to this approach relating to blank, stability, and so forth, and most probably it may be assumed that a specific activity of 100 mc/mmole is the upper practical limit (that is, 10^{-13} moles "limit sensitivity").

A second possible way involves the preparation of labeled derivatives of labeled enzyme products. To illustrate this point, consider the reaction for ethanolamine phosphokinase (Fig. 5d). After the enzyme reaction is completed both the labeled ethanolamine and ethanolamine phosphate may be methylated with C^{14}-methyl iodide to give highly labeled choline and choline phosphate, respectively. Thus, the amount of labeled choline phosphate (isolated by procedures described above) would be directly proportional to enzymatically formed ethanolamine phosphate. But the sensitivity of the procedure would be increased by the labeled methyl groups. Similarly, one could visualize the use of labeled dinitrofluorobenzene or labeled dinitrophenylhydrazine for reactions with enzymatically formed amines and aldehydes, respectively, to produce secondary products having a much higher specific activity and perhaps even more favorable properties for effective isolation. Such procedures might further increase the "limit sensitivity" to 10^{-14} to 10^{-15} moles.

Technically, there is no reason why isotopes could not be used in conjunction with the "cycling" procedures similar to those described by Lowry et al. [10]. The "cycling" procedure for NADP (NADPH) measurements is shown in Fig. 9. At the completion of the "cycle," the 6-P-gluconic is measured by the reduction of NADP in the presence of 6-P-gluconic dehydrogenase. The reaction, however, could be measured by the evolution of $C^{14}O_2$ which would occur if one had started with 1-C^{14}-glucose-6-phosphate. If isotopic techniques are to offer real utility in conjunction with "cycling" techniques, however, it would be in facilitating the development of "cycles" involving substrates and coenzymes that are not accessible to measure via the pyridine nucleotide systems. The enzyme reactions cited above, involving acetyl CoA, S-adenosylmethionine, and

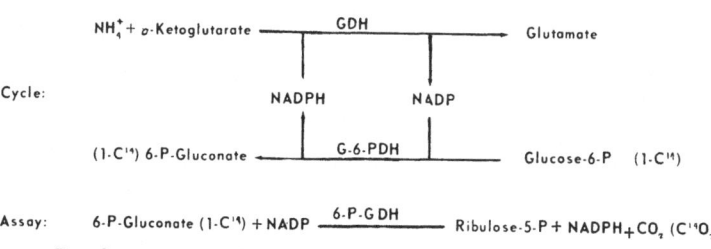

Fig. 9. Scheme for "cycling" procedure for pyridine nucleotides.

the nucleotide triphosphates, may offer some possibilities in this direction.

MICROTECHNIQUES

Since most of the quantitative isotopic procedures developed in the author's laboratories are also microtechniques, it would seem appropriate to give at least a brief consideration to the benefits of conducting such experiments on a reduced scale. Since the radioactivity is usually measured in 10 to 15 ml of counting fluid, one may reasonably ask why use μl volumes for the rest of the experiment We are obliged to work with small tissue samples (0.05 to 20 μg) and consequently small volumes (μl) in our efforts to gain biochemical information for discrete cell populations of nervous tissues. Nevertheless, there are inherent advantages to conducting the isotopic experiments on a microscale. One obvious asset is that it substantially reduces the cost of these experiments. In addition, the reduction of the quantity of isotope used in a given experiment also offers the benefit of increased safety for personnel. Perhaps not so obvious is the advantage due to the reduction of interference (quenching) of scintillation counting per se by the small volumes of aqueous solutions, organic solvents, and so on and the fact the microscale usually affords an increased efficiency in extractions and in the isolation of small precipitates. Furthermore, the small bore of the microtubes retards evaporation of volatile organic solvents and facilitates more efficient recovery of aliquots from a two-phase system.

While micropipettes, microtubes, and accessories commensurate with the scale of analysis described in detail above (Figs. 1 to 5) are in "routine" use in our laboratory, their fabrication and use generally requires much practice (and patience). However, conversion to a microscale, somewhat comparable to that under discussion, is relatively convenient due to the recent availability of well-designed, precisely calibrated, and reasonably priced micropipettes in a complete range from 1 to 500 μl (H.E. Pedersen, Sommerstedgade, Copenhagen V, Demark). A convenient size microtube compatible with these commercial pipettes may be quickly produced in large numbers by inexperienced laboratory personnel. The tubes are made from 7 mm OD pyrex tubing as previously described [28].   The resultant scale of analysis will be only about 3 times larger than that indicated for the various protocols described above.

In conclusion, it is appropriate to reemphasize the advantages of the isotopic procedures described above: (1) The techniques for measuring isotopically labeled materials have been shown to have an inherent sensitivity equal to or greater than that of other physical techniques (that is, absorbance, fluorescence, and so on) currently used for biochemical studies; (2) isotopically labeled materials have been shown to be useful for the quantitative measurement of the activities of a variety of enzymes and of the endogenous levels of various metabolic intermediates in crude tissues; (3) for some of the examples cited, the isotopic methods represent the only way that a reliable measure of the level of an enzyme or tissue constituent may be obtained; (4) the isotopic methods are frequently unaffected by secondary or side reactions which lead to gross errors in measurement when other types of methods are used; (5) isotopic methods frequently remove the burdensome necessity of using highly purified chemicals or enzymes as auxillary reagents that are required for other forms of analysis (that is, fluorescence); (6) a peculiar advantage of isotopic methods is the ease with which they may be used to study the effects of various antimetabolites or inhibitors; (7) finally, it may be so obvious that its full significance is not generally appreciated that an isotopically labeled form of a biochemical intermediate is far more of a "natural" substrate for enzyme studies than one artifically selected or synthesized for its particular chromogenic or fluorogenic properties.

It is hoped that the above considerations will lead to a better appreciation of the great potential of isotopic techniques to quantitative histochemical and cytochemical studies.

REFERENCES

1. Lowry, O.H., in Waelsch, H. (editor), Biochemistry of the Developing Nervous System, Academic Press, New York, 1954.
2. Lowry, O.H., in Cohen, M., and Snider, R. (editors), Morphological and Biochemical Correlates of Neural Activity, Harper & Row, New York, 1964.
3. Robins, E., Exp. Cell Res. Suppl. 4:241 (1957).
4. Robins, E., Smith, D.E., Eydt, K.M., and McCaman, R.E., J. Neurochem. 1:68 (1956).
5. Lowry, O.H., Roberts, N.R., Wu, M-L., Hixon, W.S., and Crawford, E.J., J. Biol. Chem. 207:19 (1954).

6. Robins, E., Smith, D.E. and McCaman, R.E., J. Biol. Chem. 204:927 (1953).
7. Glick, D., in Glick, D. (editor), Quantitative Chemical Techniques of Histo- and Cytochemistry, John Wiley, New York, 1961.
8. Lowry, O.H., Roberts, N.R., and Kapphahn, J.I., J.Biol. Chem. 224:1047 (1957).
9. Lowry, O.H., Passonneau, J.V., Hasselberger, F.X., and Schulz, D.W., J. Biol. Chem. 239:18 (1964).
10. Lowry, O.H., Passonneau, J.V., Schulz, D.W., and Rock, M.K., J. Biol. Chem. 236:2746 (1961).
11. McCaman, R.E., McCaman, M.W., Hunt, J.M., and Smith, M.S., J. Neurochem. 12:15 (1965).
12. Axelrod, J., Albers, W., and Clemente, C.D., J. Neurochem. 5:68 (1959).
13. McCaman, R.E., Life Sci. 4:2353 (1965).
14. Axelrod, J., and Tomchick, R.J., J. Biol. Chem. 233:702 (1958).
15. McCaman, R.E., and Cook, K., J. Biol. Chem. 241:3390 (1966).
16. McCaman, R.E., and Cook, K., in preparation.
17. McCaman, R.E., in preparation.
18. McCaman, M.W., and McCaman, R.E., in preparation.
19. Albers, R.W., and Brady, R.O., J. Biol. Chem. 234:926 (1959).
20. Pastan, I., Wills, V., Herring, B., and Field, J.B., J. Biol. Chem. 238:3362 (1963).
21. Schuberth, J., Acta Chem. Scand. 17, Suppl. 1:S233 (1963).
22. McCaman, R.E., and Hunt, J.M., J. Neurochem., 12:253 (1965).
23. McCaman, R.E., McCaman, M.W., and Stafford, M.L., J. Biol. Chem. 241:930 (1966).
24. Fuller, R.W., and Hunt, J.M., Anal. Biochem. 16:349 (1966).
25. Baldessarini, R.J., and Kopin, I.J., Anal. Biochem. 6:289 (1963).
26. Baldessarini, R.J., and Kopin, I.J., J. Neurochem. 13:769 (1966).
27. Snyder, S.H., Baldessarini, R.J., and Axelrod, J., J. Pharmacol. Exp. Therap. 153:544 (1966).
28. Lowry, O.H., Roberts, N.R., and M-L.W. Chang, J. Biol. Chem. 222:97 (1956).

APPLICATION OF TRACER METHODOLOGY
TO THE STUDY OF BIOGENIC AMINES

C. R. Creveling

NIAMD
National Institutes of Health
Bethesda, Maryland

The availability of isotopically labeled biogenic amines, such as norepinephrine, epinephrine, dopamine, serotonin, and histamine, has greatly increased the scope and ease of investigations of amine uptake, binding, release, biosynthesis, and metabolism. Such amines are now available with specific activities high enough (5 to 10 c/mmole) to permit studies with amine concentrations within the physiological range. This is particularly advantageous since studies now can be carried out undistorted by the gross pharmacological changes which accompany the use of larger amounts of nonisotopic biogenic amines.

One of the most successful applications of tracer methodology to the study of biogenic amines has been the study of uptake and subsequent release of exogenous catechol amines by sympathetically innervated tissues. In 1959 Axelrod et al. [1] reported that following the intravenous administration of H^3-epinephrine to mice the amine disappeared in two phases. Initially, there was a rapid metabolism of approximately 70% of the administered amine by O-methylation. The remaining 30% disappeared slowly, following accumulation in such tissues as heart, spleen, lung, and kidney. Similar experiments were performed with H^3-norepinephrine (H^3-NE), revealing an even more striking accumulation or "physiological inactivation" of the injected amine [2]. Subsequent investigations in several laboratories have indicated that uptake and retention of H^3-NE occurs predominately at postganglionic sympathetic nerve terminals. Two distinct processes appear to be operative: an initial uptake of catecholamines into the neuron followed by retention of the accumulated amines at specific intracellular

storage sites. Recent reviews have described various aspects
of the study of biogenic amines in which isotopic methodology
has played an important role [3−6].

The uptake and retention of intravenously administered
H3-NE in the mouse heart is shown in Fig. 1. The isotopically
labeled material present in the heart is due exclusively to
H3-NE, as the metabolites of NE are not retained. The second
or slow phase of H3-NE disappearance represents the sum of
normal transmitter outflow and metabolism. This is con-
trasted with the time course of H3-isoproterenol disappearance
(Fig. 1); here the cardiac level nearly parallels the plasma
level, indicating little retention. The retention of H3-NE within

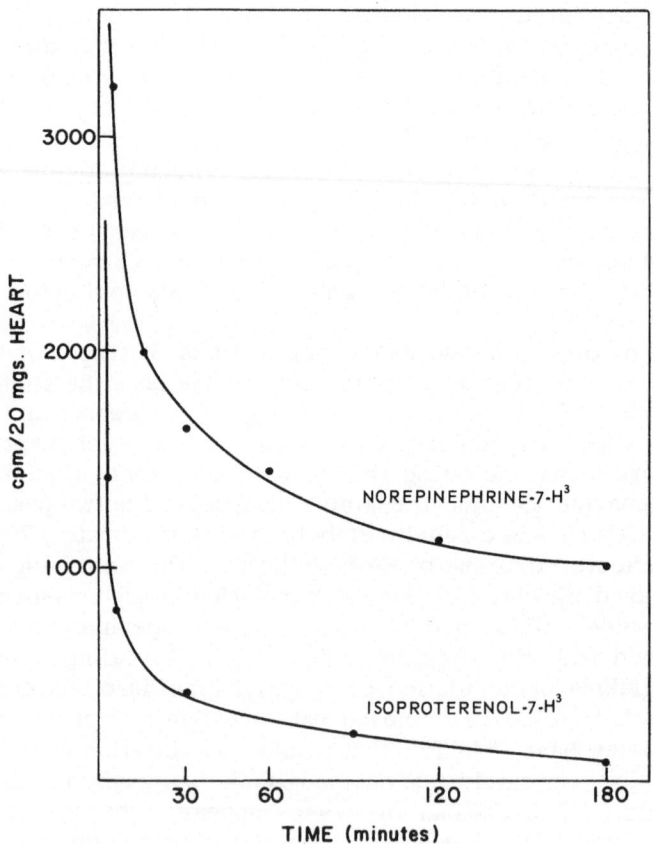

Fig. 1. Disappearance of catecholamines from heart.

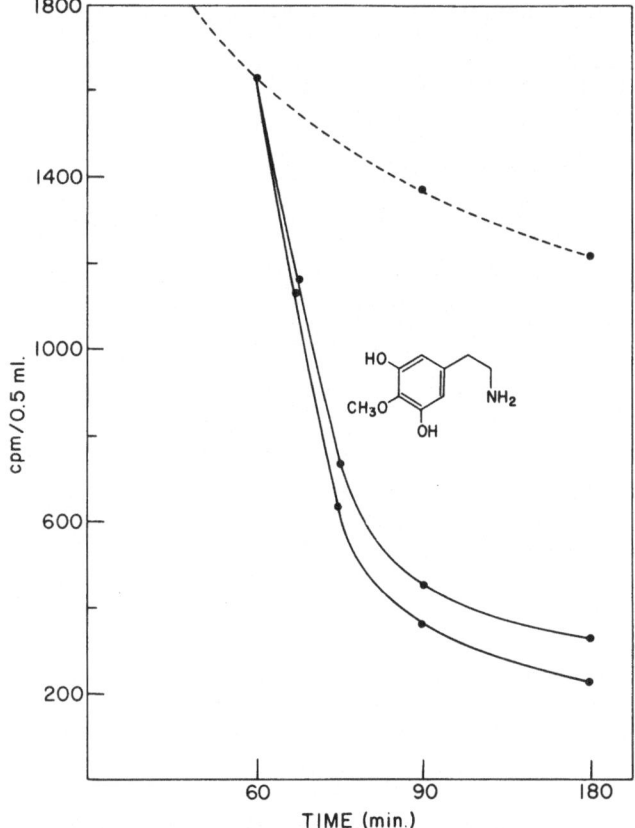

Fig. 2. Chemorelease of cardiac norepinephrine.

the heart provided a system for investigating the release phenomena, as illustrated in Fig. 2. Many compounds when administered to an animal following the initial prelabeling with H^3-NE produce a rapid increase in the rate of NE-H^3 release (Fig. 2). The decrease in H^3-NE is paralleled by a similar decrease of endogenous heart NE.

A simple and rapid method, outlined in Table I, was used to systematically evaluate the structure–activity relationship of phenethylamines and chemorelease of H^3-NE [7]. Similar studies of other classes of compounds have been reported including indoles, reserpine and related alkaloids, phenothiazines, ganglionic blocking agents, guanidines, sympatho-

Table I. Technique for Screening Drugs Which Release
Norepinephrine

---

1. Norepinephrine-7-H^3 (5 μc) injected intravenously (tail vein) to mouse at
zero time

2. Drug administered subcutaneously at 60 min

3. Animal sacrificed at 180 min; heart removed and homogenized in cold 0.4 N
perchloric acid

4. Homogenate centrifuged (10,000 × G, 10 min.); radioactivity measured in the
supernatant fluid by liquid scintillation counting.

---

lytics, and inhibitors of the norepinephrine-biosynthetic and metabolic enzymes [8, 9]. Some examples illustrating the stereospecificity and structural limitations of chemoreleasing agents are presented in Table II.

Several lines of investigation have evolved from these initial structure–activity studies relating to the mechanism of NE release and to new classes of chemoreleasers. One such discovery came from a study of the chemorelease or rather the consistent lack of chemorelease-activity produced by O-methyl-phenethylamine derivatives. It has been known for some time that O-methylation in vivo catalyzed by catechol-O-methyl

Table II. Comparison of Release of Endogenous Cardiac Norepinephrine and Norepinephrine-H^3

Drug	Dose (mg/kg)	Percent Endogenous Ne	Percent NE-H^3
Control		100*	100†
Tyramine	5.0	57	50
Metaraminol	2.5	19	14
6-OH-dopamine	5.0	40	18
3,4,5-tri-OH-PEA	5.0	45	21
3,5-di-OH-4-OCH$_3$-PEA	2.5	38	18
Reserpine	1.0	8	16
Guanethidine	10.0	29	37
Segontin	10.0	68	47
Oxypertine	10.0	45	39

*0.93 ± .16μ g/g heart.
†340 mμ c/g heart.

transferase abolishes the physiological activity of catecholamines [10]. In accordance with these observations, 3-O-methyl
dopamine, metanephrine, and normetanephrine were inactive
as chemoreleasers. The products of enzymatic O-methylation of
6-hydroxydopamine and 2,3-dihydroxycatecholamines were also
inactive. Further investigation revealed that all the O-methyl
derivatives of mono-, di-, and tri-hydroxyphenethylamines
were inactive with one notable exception, 3,5-dihydroxy-4-
methoxyphenethylamine (Table III). This compound is formed
by the action of catechol-O-methyl transferase from 3,4,5-
trihydroxyphenethylamine [11] and is considerably more active
(ED_{50}; 1.4 μmoles/kg) than its precursor (ED_{50}; 11.5 μm/kg)
[9]. Thus, it appears that the unusual activity of the 3,5-
dihydroxy-4-methoxy derivative is due to its specific structural configuration which may orient the "hydroxyl" groups in
a position favorable to binding at the norepinephrine storage
site. The norepinephrine analog, 3,5-dihydroxy-4-methoxy-
phenethanolamine (ED_{50}; 5.0 μmoles/kg) is less active as a

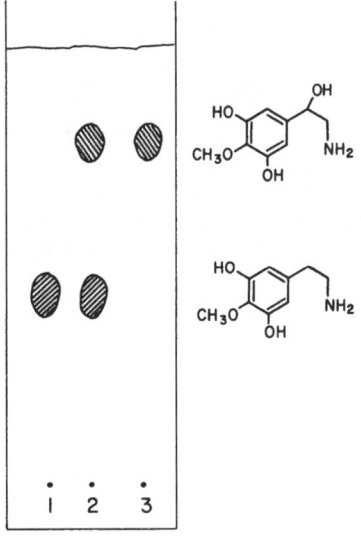

1) HEATED ENZ.   2) REACTION MIXTURE
3) AUTHENTIC PRODUCT

Fig. 3. Dopamine-β-hydroxylase: action on 3,5-dihydroxy-4-methoxy phenethylamine.

Table III. Effect of O-Methyl Substitution on Chemo-
Release*

Parent phenol	Percent release	O-methyl derivative	Percent release
Tyramine	52	4-OCH$_3$	0
Dopamine	50	3-; 3,4-	0; 0
L-norepinephrine	88	3-; 4-	0; 0
2,3-dihydroxy PEA	57	3-	0
3,5-dihydroxy PEA	50	3-	0
2,4,5-trihydroxy PEA	72	2-; 4-; 5-	0,10,0
3,4,5-trihydroxy PEA	88	3-; 3,4-; 3,5-	5,0,0
		4-	88

*Dose: 5 mg/kg/

releasing agent. This may result from a less rapid uptake into
the heart since phenethylamines are frequently transported
more readily into sympathetic nerve than their phenethanolamine analogs [12, 13].

Many phenethylamines are substrates for dopamine-β-
hydroxylase both in vitro [14] and in vivo [15]. Amines such
as tyramine are taken up from the circulation into sympathetic
neurons, undergo β-hydroxylation, and displace norepinephrine
from intraneuronal storage sites. Such norepinephrine analogs
can then be released by nerve stimulation in a manner similar
to the normal transmitter and have been termed "false trans-
mitters" [16, 17]. It has been shown that this rather unusual
norepinephrine releaser, 3,5-dihydroxy-4-methoxyphenethyl-
amine, is not only a substrate for purified dopamine-β-hydrox-
ylase (Fig. 3) but it also is taken up into heart tissue and undergoes β-hydroxylation in vivo. This was demonstrated through
the isolation of the injected phenethylamine and its β-hydrox-
ylated product from mouse heart as the highly fluorescent
1-dimethylaminonaphthalene-5-sulfonyl derivative (dansyl derivative) (Fig. 4). This method has been used to study the
amines normally present in cardiac tissue and to follow the
uptake and retention of various exogenous administered amines.
Identification of the isolated dansyl amines is accomplished
by fluorometry, isotopic dilution, and mass spectrometry [18].
While all the criteria for establishing the 3,5-dihydroxy-4-
methoxy-derivative as a false transmitter have not been satis-
fied, such a role is indicated.

The interaction between blocking agents such as cocaine and

bretylium and norepinephrine-releasing agents can be studied using the technique of prelabeling cardiac stores of NE with H^3-NE. Cocaine is believed to act at the neural membrane, blocking the uptake of simple amines such as tyramine [19]. Bretylium blocks the release of catecholamines by reserpine and related compounds and probably interacts interneuronally, possibly at the storage granule [20]. A comparison was made of the effect of cocaine and bretylium on the release of norepinephrine by 3,5-dihydroxy-4-methoxyphenethylamine and related compounds. As shown in Fig. 5, neither cocaine nor bretylium were effective in blocking norepinephrine release by 3,4,5-trihydroxyphenethylamine or metaraminol. With the introduction of a 4-methoxy group to form the 3,5-dihydroxy-4-methoxy derivative, release now can be blocked by bretylium as can release by reserpine and tetrabenazine. Release by the β-hydroxylated derivative is blocked both by cocaine and bretylium similar to the effect on release by 6-hydroxydopamine (2,4,5-trihydroxyphenethylamine) [21]. Certain amines such as 6-hydroxydopamine produce prolonged depletion of cardiac norepinephrine [22]. Reserpine and metaraminol also cause

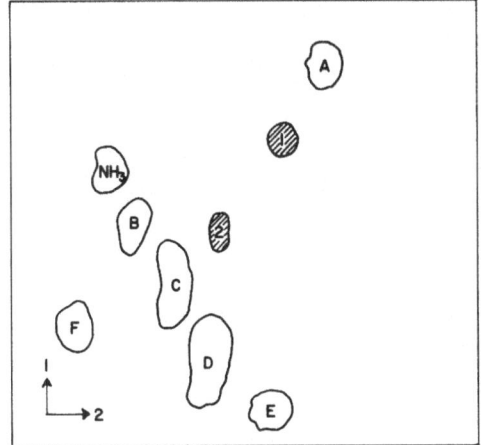

A–F · NORMALLY OCCURRING AMINES
I, 2 · APPEAR AFTER I.V. INJECTION (IO MGS/KG)

Fig. 4. Cardiac retention of 3,5-dihydroxy-5-methoxyphenethylamine: TLC of dansyl derivatives from mouse heart.

prolonged norepinephrine depletion. To ascertain the duration of the impairment of norepinephrine uptake and storage mechanisms, the uptake and release of H^3-NE at varying intervals after administration of drug were studied. As shown in Fig. 6 the effect of 3,5-dihydroxy-4-methoxyphenethylamine on H^3-NE uptake and release had disappeared after 24 hr while the 3,4,5-trihydroxy derivative was still moderately effective.

Despite the rapid and complete depletion of cardiac norepinephrine produced by 3,5-dihydroxy-4-methoxyphenethylamine, little gross sympathomimetic activity was observed in the mouse. In addition, no increase in heart rate with concentrations of this amine up to 5×10^{-4} M was observed in the isolated, perfused rat heart. The pressor activity of this amine in the intact, anesthetized rat was minimal, approximately 1/200 that of norepinephrine [23]. One possible explanation consistent with these observations would be an interaction of 3,5-dihydroxy-4-methoxyphenethylamine with the norepinephrine storage granule resulting in intraneuronal release and

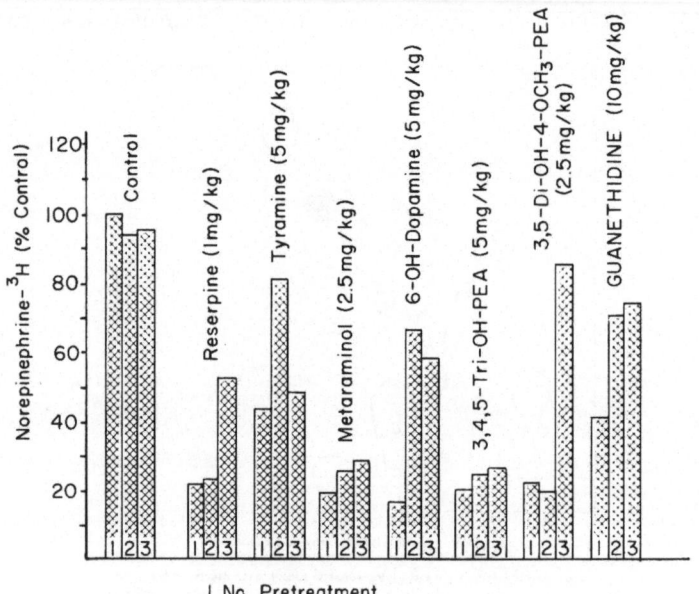

1 No Pretreatment
2 Pretreatment Cocaine 20 mg/kg
3 Pretreatment Bretylium 20 mg/kg

Fig. 5

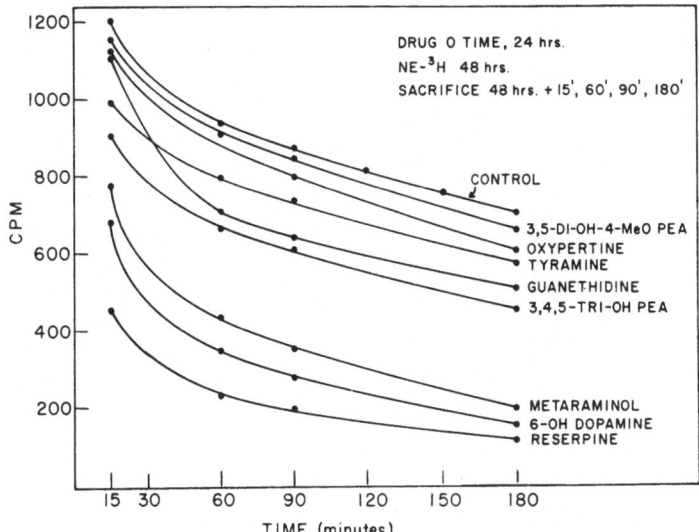

Fig. 6. Long-term effect of drugs on norepinephrine release.

exposure of unbound transmitter to metabolism by intraneuronal monoamine oxidase. Such a mechanism has been proposed for the alteration of metabolic products from norepinephrine following reserpine-induced release [24]. Reserpine-induced release results in a decrease in normetanephrine and an increase in deaminated products.

To explore this possible mechanism of release, metabolic studies were carried out in the isolated perfused rat heart. The pattern of metabolic products derived from H^3-NE appearing in the rat heart perfusate after treatment with reserpine, tyramine, and 3,5-dihydroxy-4-methoxyphenethylamine are shown in Table IV. The changes in metabolic pattern following reserpine were minimal; however, a small decrease in normetanephrine and measurable quantities of catechol acids were observed. These observations are in agreement with the suggestion of Hertting [20] that effective release by reserpine requires sympathetic innervation of the heart. Tyramine, on the other hand, produced a significant increase in normetanephrine and concomitant decrease in VMA, indicating an increased participation of catechol-O-methyl transferase activity. This is interpreted as similar to the normal, neurogenic release of transmitter at the

Table IV. Release and Metabolism of Norepinephrine-H^3 in the Isolated-Perfused Rat Heart

Drug	Concentration	Percent R	Catechol	Non-catechol	NE	Catechol acids	NMN	VMA
Control		14	4.0	96	4.0	1.0	22	74
Reserpine	3×10^{-5} M	14	12.0	88	9.5	2.5	17	71
Tyramine	5×10^{-4} M	14	12.5	88	9.1	3.4	59	28
3,5-dihydroxy-4-methoxy PEA	5×10^{-5} M	18	27.0	73	22.0	5.0	18	55
The same	1×10^{-4} M	23	23.0	77	19.0	4.3	25	52
The same	5×10^{-4} M	31	35.0	65	28.0	6.6	19	46

synapse where norepinephrine has direct access to catechol-O-methyl transferase on the surface of the receptor. The pattern obtained following release by 3,5-dihydroxy-4-methoxyphenethylamine was not similar to that obtained with reserpine or tyramine. There was a substantial increase in catechol acids, a moderate decrease in VMA, and essentially no change in the normetanephrine fraction. However, there was a marked increase in the amount of free norepinephrine released. This rather puzzling observation is currently under investigation and has been materially aided through the use of isotopically labeled 3,5-dihydroxy-4-methoxyphenethylamine (prepared by the New England Nuclear Corporation). Preliminary results suggest that this amine interacts not only with norepinephrine storage sites in peripheral sympathetic nerve terminals but also at adrenergic receptors.

REFERENCES

1. Axelrod, J., Weil-Malherbe, H., and Tomchick, R., J. Pharmacol. 127:251 (1959).
2. Whitby, L.G., Axelrod, J., and Weil-Malherbe, H., J. Pharmacol. 132:193 (1961).
3. Shore, P.A., Pharmacol. Rev. 14:531 (1963).
4. Zaimis, E., Ann. Rev. Pharmacol. 4:365 (1964).
5. Iversen, L.L., in Harper, N.J., and Simmonds, A.B. (editors), Advances in Drug Research, Vol. II, Academic Press, London and New York, 1965.
6. Glowinski, J., and Axelrod, J., Pharmacol. Rev. 18:775 (1966).
7. Daly, J.W., Creveling, C.R., and Witkop, B., J. Med. Chem. 9:273 (1966).
8. Daly, J.W., Creveling, C.R., and Witkop, B., J. Med. Chem. 9:280 (1966).
9. Creveling, C.R., Daly, J.W., and Witkop, B., J. Med. Chem. 9:284 (1966).
10. Bacq, Z. M., and Renson, J., Bull. Acad. Roy. Med. Belg. 25:755 (1966).
11. Daly, J.W., Axelrod, J., and Witkop, B., Ann. N.Y. Acad. Sci. 96:37 (1962).
12. Iversen, L.L., Brit. J. Pharmacol. 25:18 (1965).
13. Burgen, A.S.V., and Iversen, L. L., Brit. J. Pharmacol. 25:34 (1965).
14. Creveling, C.R., Daly, J.W., Witkop, B., and Udenfriend, S., Biochim. Biophys. Acta. 64:125 (1962).
15. Musacchio, J., Kopin, I.J., and Weise, V.K., J. Pharmacol. Exp. Therap. 148:22 (1965).
16. Fischer, J.E., Horst, W.D., and Kopin, I.J., Brit. J. Pharmacol. 24:477 (1965).
17. Cohen, R.A., Ann. Intern. Med. 65:347 (1966).
18. Daly, J.W., Kondo, K., and Creveling, C.R., in preparation.
19. Trendelenberg, U., J. Pharmacol. 125:55 (1959).
20. Hertting, G., Axelrod, J., and Patrick, R.W., Brit. J. Pharmacol. 18:161 (1962).
21. Creveling, C.R., Daly, J.W., and Witkop, B., in preparation.
22. Porter, C.C., Totaro, J.A., and Stone, C.A., J. Pharmacol. Exp. Therap. 140:308 (1963).
23. Spector, S., personal communication.
24. Lindmar, R., and Muschol, E., Arch. Exp. Pathol. Pharmakol. 247:469 (1964).

USE OF CHRONICALLY LABELED ANIMALS IN THE STUDY OF A "METABOLICALLY INERT" PROTEIN

LeRoy Klein

Department of Surgery
Division of Orthopedic Surgery
Western Reserve University School of Medicine
Cleveland, Ohio

The use of acute labeling techniques (pulse labeling) has been successful in tracing and locating rapid metabolic pathways *in vivo*. This procedure was used in the classical experiments of Schoenheimer [1] on labile cellular proteins which gave rise to the concept of the dynamic state of body constituents. The concept described protein turnover as a continual process of synthesis and degradation as evidenced by the active uptake and release of isotopic compounds (see Fig. 1). Turnover represented the occurrence of synthesis which cannot be detected analytically because it was exactly balanced by breakdown [2]. Acute and chronic isotopic studies have demonstrated that the fibrous proteins of muscle and connective tissues have the least active turnover and are considered to be relatively inert metabolically [3, 4]—inert in the sense that the proteins were nonrenewable. However, these studies could not distinguish between essentially *non*isotopic turnover (nonsynthetic and nondestructive) of macromolecules and absolute inertness.

The major limitations of pulse labeling *in vivo* have been in quantitating metabolic pathways. These problems are particularly evident in the higher organisms, where there is both anatomical complexity (circulatory system and multiple permeability barriers) and metabolic heterogeneity of tissues

This investigation was supported in part by U.S. Public Health Service institutional grants HD-00669 and FR-5410 from The Institutes of Child Health and Human Development, and General Medical Sciences, and in part by the Rainbow Hospital Research Fund.

215

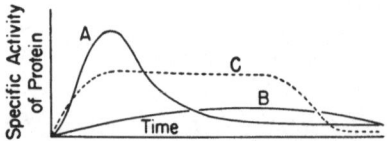

Fig. 1. General types of specific activity—time curves obtained following the administration of a single dose of labeled amino acid. Reprinted with permission from H. Tarver [6].

[5]. This complexity results in a slow equilibrium of metabolic pools, reversibility of reactions, and recycling of metabolites. Another difficulty in the interpretation of protein turnover *in vio* is related to the comparison of synthetic rates without consideration of the pool size of the preexisting molecules. When the pool size of preexisting proteins becomes large in comparison to that of the newly synthesized proteins, the significance of low specific activities as a measure of turnover becomes limited [6]. The converse of this situation is seen in the embryo where the pool size of preexisting molecules is small. By the proper choice of animal age and experimental design for administering isotope [5] one could estimate the quantitative significance of one reaction to the sum total of metabolic pathways.

Since turnover can be understood as being either the synthesis of new molecules or the new arrival of preexisting molecules [2], these two possibilities need to be examined theoretically and experimentally. The first pathway represents a synthetic pathway which can be easily seen isotopically, and the second pathway represents a nonsynthetic pathway which can be visualized in the transport of soluble proteins. Studies on the reutilization of soluble proteins were inconclusive due to the inability to distinguish between protein biosynthesis and transport of protein. Conceptually the reutilization of soluble proteins was usually considered in terms of a synthetic process. However, turnover of macromolecules like DNA, collagen, myosin, and myelin, which are a part of morphological structures, could occur by nonisotopic transformation mechanisms which involve the dissociation of supermolecular structures to a molecular state and its eventual reassociation.

The testing of this possibility either by pulse or continuous labeling procedures is complicated by the reutilization of amino acids. Historically, the quantitative aspects of relabeling have been exaggerated by the use of N^{15} because re-

utilization was on a considerably greater scale due to trans-
amination and a greater animal conservation of nitrogen than
of carbon [7]; there also appeared to be a greater conservation
of carbon than of hydrogen [4].

The study of the metabolic fate of existing collagen requires that the experimental animal be in a steady state. The
growing animal is in a nonsteady state biologically [1] and pulse
labelling is a nonsteady state isotopically [5]. The use of rats
which can be repeatedly labeled with H^3-proline during their
active growth period (4 to 9 weeks of age) followed by 1 to 7
months of maturing without labeling would permit an approxi-
mation of the steady state both biologically and isotopically.
The maturing period after labeling allows for the catabolism
of labile collagen [8] and noncollagenous proteins, and thus reduces the reutilization of radioactive amino acids.

Collagen as an insoluble protein system circumvents some
of the experimental problems that have existed in the study of
reutilization of soluble proteins [9]. The formation of connective tissue (collagen) by polyvinyl sponge implantation or
carrageenan inoculation [10] has the advantages of inducing new
tissue formation in an adult animal and permitting the separation of new tissues from old radioactive areas. The chromatographic isolation and quantitation of hydroxyproline is used
to determine collagen content since nearly all of the hydroxyproline in mammals is found in collagen. A comparison of
specific activities of hydroxyproline and proline from normal
and induced connective tissues allows each animal to be his
own control.

MATERIALS AND METHODS

Weanling female or male Fischer rats weighing 45 to 50 g
were placed on Purina rat chow and water ad libitum. During
the active growing phase, each animal was given an intraperitoneal injection of L-proline-3, 4-H^3 (New England Nuclear
Corp., 5 c/mM) twice per week for 3 to 5 weeks [10]. The dose
of radioactive proline was 0.25 μc/g body weight or a total of
200 to 260 μc of proline-H^3/animal. The animals were used
experimentally 1 to 7 months after the last injection of proline-
H^3. New collagen formation was induced by implanting sterile
polyvinyl sponges [10] under the skin of labeled rats. The
methods for removal of sponge, extraction and purification of

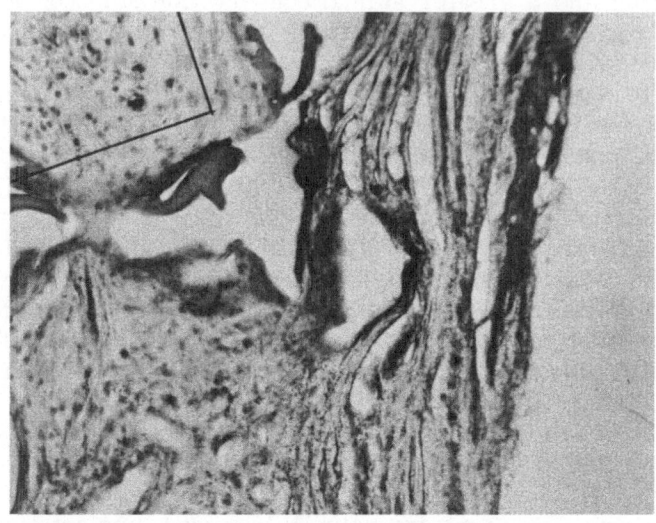

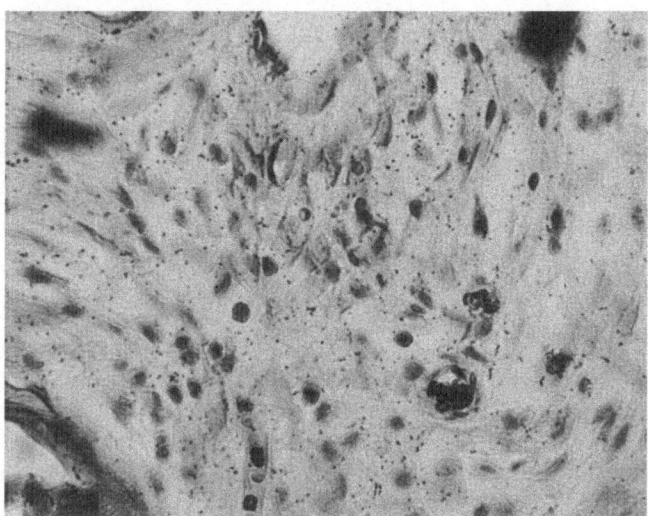

Fig. 2. Radioautographs of granulation tissue in a polyvinyl sponge after 3 weeks of implantation in a rat prelabeled with C^{14}-proline one month prior to implantation. Upper: peripheral area, × 130; lower: more central area of sponge, × 550; orientation shown in marked area of upper figure. Radioautographs exposed for 4 weeks, stained by hematoxylin and eosin.

collagen, column chromatography of amino acids, and its chemical and radioactive analysis are given elsewhere [10].

Ideally, the goal is for maximal incorporation of isotope during the period of rapid growth [5] and continuous labeling until the animal stops growing. The most suitable animal appears to be the female Fischer rat since as an adult she has a much slower growth rate than her male counterpart, and generally the adult Fischer animal grows slower than adult Sprague-Dawley or Wistar rats. In addition they are isogenic, resulting in less biological variation.

Since collagen forms insoluble fibrous layers, an additional goal is to obtain a uniform distribution of radioactivity in collagens of all ages by minimizing the amount of nonradioactive collagen added on by growth once labeling has stopped. Moreover, a sufficient period of time is needed after labeling to permit: (1) destruction of exogenous radioactive proline, (2) catabolism of radioactivity in soluble collagen, parenchymal, and plasma proteins, and (3) maturation and insolubilization of collagen and muscle proteins.

RESULTS AND DISCUSSION

I. Distribution of Radioactivity in Sponge

Information about distribution of radioactivity in space and time will aid in delineating contamination, reutilization of radioactive amino acids, and reutilization of collagen. Figure 2 is a radioautograph of granulation tissue in a polyvinyl sponge after 3 weeks of implantation in a prelabeled rat. The more central areas demonstrate an even distribution of radioactivity. The edge of the sponge has a dense fibrous capsule which appears more radioactive than the central areas. The specific activity of hydroxyproline isolated from the capsule was 2.5 to 4 times greater than that isolated from the interior of the sponge [10]. Since the origin of the capsular material was not certain, in all subsequent experiments the capsule was completely stripped away.

The degree of physical contamination that occurs by placing a sponge for 2 weeks in a prelabeled animal can be estimated by simultaneously placing the animal on high dosages of corticoid hormones [11] which prevent vascular, cellular, and

Table I. Effect of High
Dosages of Steroid* upon
Collagen Content of Im-
planted Sponge †

Rat number	Collagen (mg)
8	0.15
13	0.34
15	0.32
20	0.13
21	0.10
22	0.19
23	0.10
Average	0.19

*Dexamethasone (0.25 mg) and tri-
amcinolone (0.5 mg) administered
directly into sponge at time of implantation and 1 week later.
†Sponges implanted into male Fischer rats for 2 weeks, 4 months
after labeling.

collagenous infiltration of the sponge. Approximately 0.2 mg
of collagen was found within the sponge (see Table I) as compared to the normal range of 10 to 20 mg of collagen. This
represented a relatively small amount of radioactivity. In
addition, no adhering capsule formed around the sponge after
hormonal treatment. The absence of the fibrous capsule in
steriod-treated rats suggested that the capsule in untreated
animals was the result of active growth rather than physical
condensation of loose connective tissues around the periphery.

In brief, the sponge induced an adequate amount of "new"
collagen which could be separated from the surrounding radioactive tissues with minimal contamination. The radioactive
protein found within the sponge has the physical and chemical
properties of collagen, as described elsewhere [10].

II. Distribution of Radioactivity in the Animal

Since collagen and muscle proteins make up to 80 to 90%
of the total protein in the body [12], these two proteins represent the major pool of radioactive proline as synthetic pre-

Table II. Specific H³ Activities of Proline Isolated from
Proteins of Chronically Labeled Rats

Months after labeling	Collagen (dpm/µg proline)			Noncollagenous protein (dpm/µg proline)	
	Bone	Skin	Sponge*	Sponge*	Muscle
2†	18.8	8.3	4.1	2.4	6.4
	(226)‡	(100)	(49)	(29)	(77)
7§	23.8	8.7	2.7	0.43	2.7
	(274)	(100)	(31)	(5)	(31)

*Sponges were implanted for 2 weeks.
†Two male and two female Fischer rats received 215 µc of H³-proline in
nine injections over 3 weeks.
‡Figures in parentheses represent a percentage of skin specific activity.
Noncollagenous proteins of sponge and muscle were prepared by treating
3 times in 5.5% trichloracetic acid at 90° C for 30 min.
§Two male and four female Fischer rats received 255 µc of H³-proline
in ten injections over 5 weeks.

cursors for hydroxyproline. In rats the highest specific activities of proline were found in bone collagen, followed by skin collagen and muscle protein (see Table II). Two months after labeling, proline specific activity of sponge collagen was one half of that of skin collagen and higher than sponge noncollagenous protein (mostly serum proteins). These observations in rats 7 months after labeling were somewhat similar with the exception of sponge noncollagenous protein which was one sixth that of skin collagen. The proline specific activity of sponge collagen and muscle proteins was one third that of skin collagen. It appears that any long-term reutilization of

Table III. Proline-H³ and Hydroxyproline-H³ as a Percentage
of Total Radioactivity in Protein

Months after labeling	Collagen (proline + hydroxyproline)			Noncollagenous protein (proline)	
	Bone (percent)	Skin (percent)	Sponge* (percent)	Sponge* (percent)	Muscle (percent)
2	90	88	77	72	76
7	95	94	79	–	78

*See footnotes for Table II.

Table IV. C^{14}-Amino Acids as a Percentage of Total
Radioactivity in Proteins

Animal	Radioactive amino acid	Percentage of protein radioactivity	References
Rat *in vivo*	DL-tyrosine-C^{14}	62−90	[13]
Rat *in vivo*	glycine-C^{14}	60−68	[14]
Rat liver slices	DL-alanine-C^{14}	92−100	[15]
Rat liver homogenate	leucine-C^{14}	90−100	[16]
Guinea pig liver homogenate	lysine-C^{14}	92−95	[17]

radioactive free or protein-bound proline would be derived more likely from collagen than muscle proteins to account for the high specific activity in sponge collagen.

To determine how much radioactivity spills over into other amino acids in chronically labeled animals the sum of the radioactivity in proline and hydroxyproline in collagen or proline in noncollagenous proteins was compared to the total radioactivity in protein hydrolyzates. Table III demonstrates that proline and hydroxyproline account for 77 to 95% of the total radioactivity present in collagen hydrolyzates from normal and induced tissues, and proline accounts for 72 to 78% in the hydrolyzates of noncollagenous proteins. The collagen percentages are in agreement with the *in vitro* pulse-labeling data from the literature (see Table IV) but are higher than the *in vivo* data.

III. Reutilization of Collagen and Muscle Catabolites

The purpose of this study was to estimate how much of the total radioactivity in sponge was derived from the reutilization of protein catabolites (free proline). Chronically labeled female rats were made pregnant 1 and 4 months after labeling as a means of inducing *de novo* collagen synthesis; 1 week later a sponge was implanted within the same animals, and after 2 weeks both sponge and fetus were removed. Since nearly all of the fetal collagen is synthesized in the last 2 weeks of the rat's pregnancy, the sponge and fetal collagen were formed during the same period of time. A comparison of the specific activities of hydroxyproline isolated from the various collagens demonstrated that 1 month after labeling, fetal collagen and maternal sponge collagen were 7 to 8% and 25 to 30%, respectively, of skin collagen (unpublished observations).

One group of pregnant rats was on a regular diet while the second group was on the same diet but in addition contained 15% nonradioactive proline for 3 weeks. This latter group daily consumed 2.4 g of nonradioactive L-proline in their diet which was 10 times their normal intake. A polyvinyl sponge was implanted within each animal, and after 2 weeks both sponge and fetus were removed. The pregnant rats that were fed L-proline showed a tenfold decrease in specific activity of fetal hydroxyproline while there was little change in sponge hydroxyproline activity (see Table V).

A comparison of proline incorporation rates into the maternal sponge and fetus was made after an *acute* administration of proline-H^3 to nonradioactive pregnant rats. The fetus incorporated 5 times more radioactivity into hydroxyproline per milligram collagen than the sponge (unpublished observations). The sensitivity of the fetus in detecting radioactive amino acids in the maternal system appears to be derived from its comparative greater ability to concentrate amino acids and the low catabolism of amino acids by fetal liver [18].

Using fetal radioactivity as an estimation of the systemic reutilization of free proline the comparison of the specific activities of collagen-bound hydroxyproline in sponge and fetus indicated that radioactive proline or low molecular weight peptides were not significantly reused to label sponge collagen. Hydroxyproline per se cannot be the source of radioactivity in fetal collagen since hydroxyproline is not utilized in *de novo* collagen synthesis [19]. These studies do not rule out the possibility that collagen is destroyed intracellularly and its proline reused with little dilution. Evidence for collagen reutilization in the scorbutic state [10] tends to negate local reutilization of proline intracellularly.

Table V. Comparison of Collagen Synthesis and Collagen Reutilization in the Pregnant Rat

Condition	Diet	Maternal skin hypro sa*	Maternal sponge†		Fetus hypro sa*	Hydroxyproline relative specific activity‡	
			Collagen mg	hypro sa**		$\frac{\text{Sponge}}{\text{Skin}} \times 100$	$\frac{\text{Fetus}}{\text{Skin}} \times 100$
Pregnant	Normal	6.7	17.8	2.2	0.28	33	4.2
		6.6	16.8	1.2	0.22	19	3.3
Pregnant	15% L-proline	7.8	9.7	1.3	0.018	16	0.22
		8.0	22.4	2.3	0.029	29	0.36
Nonpregnant	15% L-proline	3.9	11.8	0.60		15	
		6.6	11.7	1.6		24	

*sa = specific activity (dpm/µg hydroxyproline),

†Four months after labeling rats with proline-H^3, sponges were implanted for 2 weeks.

‡Hydroxyproline relative specific activity = $\frac{\text{sponge hypro sa}}{\text{skin hypro sa}} \times 100.$

It is beyond the scope of this report to go further into additional evidence concerning proline reutilization, but it can be mentioned that bone formation within diffusion chambers, liver fibrosis, and parabiosis have demonstrated a relatively insignificant amount of proline reutilization in chronically labeled animals (unpublished observations).

CONCLUSION

The presented data indicate that there is a significant reutilization of preexisting radioactivity in the formation of "new" connective tissues. This reutilization appears to be distinct from the reutilization of proline by *de novo* protein synthesis. It has been suggested [10] that the reutilization of preexisting collagen occurs by the depolymerization of fibrous collagen (aggregates of macromolecules) to soluble molecular units or subunits and the subsequent extracellular polymerization to fibrous collagen. Conceptually, protein reutilization would result in no *net* change in radioactivity [20] during pulse-labeling experiments and would give the appearance of metabolic inertness while still participating in active turnover at a *super*molecular level.

The major advantage of adult animals that were chronically labeled during their period of active growth is that they approach a steady state isotopically and biologically, thus permitting a quantitative interpretation of specific activity [5, 6]. This isotopic design allows an evaluation of the fate of "metabolically inert" macromolecules like DNA, collagen, myosin, and myelin in the adult animal. It has many advantages over pulse-labeling experiments *in vivo*. Long-term studies can be used to evaluate slow metabolic pathways in the aged animal and to distinguish between destructive and nondestructive routes [21] which often cannot be duplicated by *in vitro* experiments. There is a greater flexibility of experimental design isotopically (chronic and acute labeling, double labeling) and biologically (parabiosis, induction of new tissue).

The major disadvantages of chronic labeling is the high cost of labeling and maintaining the animals, the long periods of time that are involved experimentally, and the presence of permeability barriers for macromolecules.

REFERENCES

1. Schoenheimer, R., The Dynamic State of Body Constituents, Harvard University Press, Cambridge, p. 5, 1942.
2. Reiner, J. M., Arch. Biochem. Biophys. 46:53 (1953).
3. Sprinson, D. B., and Rittenberg, D., J. Biol. Chem. 180:715 (1949).
4. Thompson, R. C., and Ballou, J. E., J. Biol. Chem. 223:795 (1956).
5. Davison, A. N., and Dobbing, J., Nature 191:844 (1961).
6. Tarver, H., in Neurath, H., and Bailey, K. (editors), The Proteins, Vol. IIB, Academic Press, New York, p. 1259, 1954.
7. McFarlane, A. S., Bull. Swiss Acad. Med. Sci., 21:173 (1965).
8. Lindstedt, S., and Prockop, D. J., J. Biol. Chem. 236:1399 (1961).
9. Campbell, P. N., Advan. Cancer Res. 5:97 (1958).
10. Klein, L., and Weiss, P. H., Proc. Natl. Acad. Sci. 56:277 (1966).
11. Michael, M., and Whorton, C. M., Proc. Soc. Exp. Biol. Med. 76:754 (1951).
12. Everett, M. R., Medical Biochemistry, P.B. Hoeber, New York, pp. 410, 570, 1948.
13. Winnick, T., Friedberg, F., and Greenberg, D. M., J. Biol. Chem. 173:189 (1948).
14. Greenberg, D. M., and Winnick, T., J. Biol. Chem. 173:199 (1948).
15. Zamecnik, P. C., Frantz, I. D., Loftfield, R. B., and Stephenson, M. L., J. Biol. Chem. 175:299 (1948).
16. Zamecnik, P. C., and Keller, E. B., J. Biol. Chem. 209:337 (1954).
17. Borsook, H., Deasy, C. L., Haagen-Smit, A. J., Keighley, G., and Lowy, P. H., J. Biol. Chem. 179:689 (1949).
18. Christensen, H. N., in Munro, H.N., and Allison, J.B. (editors), Mammalian Protein Metabolism, Vol. I, Academic Press, New York, p. 105, 1964.
19. Stetten, M. R., J. Biol. Chem. 181:31 (1949).
20. Neuberger, A., Brit. Med. Bull. 8:210 (1952).
21. Klein, L., and Weiss, P. H., in Comte, P., (editor) International Symposium on the Biochemistry and Physiology of Connective Tissue, Ormeco et Imprimerie, Lyon, p. 443, 1966.

RECENT ADVANCES IN THE DOUBLE ISOTOPE DERIVATIVE ANALYSIS OF STEROIDS

Bernard Kliman

Harvard Medical School
Boston, Massachusetts

The first description of the successful use of acetic anhydride-H^3 for double-isotope derivative analysis of aldosterone and other steroids appeared in 1960 [1]. Since then, this methodology has been modified by many laboratories and applied to a variety of steroids in diverse biological fluids. The original methods were described in previous symposia [2, 3]. The purpose of this paper is to summarize several changes in methodology and new applications of such methods in one laboratory. Other workers have developed and applied their own modifications, which are too numerous to review here. The purpose of these changes is to provide greater reliability, speed, and sensitivity for analyses of steroids in biologic samples. The general method is applied to selective assay of specific steroids, with a range of sensitivity to 0.01 μg using 100 mc/mmole acetic anhydride-H^3.

1. USE OF 4-C^{14}-STEROIDS

The recovery of each steroid in the sample is based on the recovery of the C^{14}-labeled tracer identical to the unknown. It was customary to synthesize the companion tracer from a reference supply of the steroid by acetylation with acetic anhydride-C^{14} and chromatographic purification. The standard acetate derivatives were relatively stable but only provided recovery correction beyond the acetylation step. When tritium-labeled steroids are used for this purpose, the acetylating reagent is acetic anhydride-C^{14}, which is considerably more expensive than the tritium reagent. In the past 4 years, most of the steroids of biologic interest have been provided commercially with C^{14} in the 4 position at high specific activities of over 40 mc/mmole. The 4-C^{14}-steroids have the advantage of correcting for all potential loss in the entire analytical procedure, which is now termed "double isotope derivative dilution." In some instances, particularly with aldosterone and testosterone, these tracers are more susceptible to decomposition than the acetate derivatives formerly used. For this reason we have used a purity test to determine the composition of the tracer. The usual procedure is to chromatograph the steroid with unlabeled carrier and obtain a scan in a radioactivity strip detector. This test is qualitative and does not readily provide an exact measure of the percent purity of the tracer. The purity test, as used by us, is carried out for each new supply of tracer and whenever there is any indication of tracer breakdown. The tritium-labeled acetate of the steroid is separately prepared and purified. Aliquots of 4-C^{14}-steroid are mixed with carrier and reacted with unlabeled acetic anhydride. A mixture of the acetylated C^{14}-tracer and of the H^3-acetate-standard is then migrated in at least two chromatographic systems. In the case of aldosterone, the 18,21-diacetate is also oxidized to the 21-acetate, 11, 18-lactone. The yield of H^3 steroid is used to calculate initial C^{14} steroid of the same identity and thereby provides a value for percent purity of the tracer. An alternate procedure is to use 1,2-tritium-labeled steroid which is of known purity and likewise form and purify the acetate derivatives of the two tracers.

One disadvantage of the 4-C^{14}-steroid tracer is that it is measured by the tritium acetylation procedure. It is possible to use only 0.005 μg of tracer in most experiments, and in certain procedures, such as metabolic clearance rate studies, as little as 0.0005 μg. Aliquots of the tracer must be assayed by tritium acetylation in order to determine the exact contribution of the tracer. When the specific activity of a given tracer is well established, the tritium:C^{14} ratio obtained is a measure of the specific activity of the tritium reagent.

The 4-C^{14}-tracer, as used in the double-isotope derivative analysis, is particularly useful in the assay of whole blood and tissues since appreciable losses result during the extraction and alkali wash procedures which are required to obtain a low residue sample for assay. An example is the collection of blood from the left adrenal vein of intact, anesthetized rats.

The whole blood is mixed with heparin solution containing $4\text{-}C^{14}$-aldosterone and $4\text{-}C^{14}$-corticosterone. To compensate for the 100 times larger amount of corticosterone, only one tenth of the sample is assayed for this steroid, and the tritium reagent is diluted to 10 mc/mmole. After serial 10 to 15 min collections of blood are obtained, the adrenal gland is homogenized in 20 percent ethanol in saline containing similar tracers. The two groups of analyses measure adrenal vein output, adrenal gland content, and turnover rate of the steroid pool in the gland. The results of a typical experiment are shown in Fig. 1. Groups of five animals were used as sham injected controls or treated for 1 to 2 weeks with 250 μg/day angiotensin-amide sc. in oil [4]. After 2 weeks there was a marked increase in the secretion rate and adrenal gland content of both aldosterone and corticosterone. The relative turnover rate of the adrenal steroid pool was 8.2%/min, control, and 4.8%/min after 14 days for aldosterone; and 11%/min, control, and 20%/min after 14 days for corticosterone. This approach is useful in comparisons of turnover rate at varying rates of secretion.

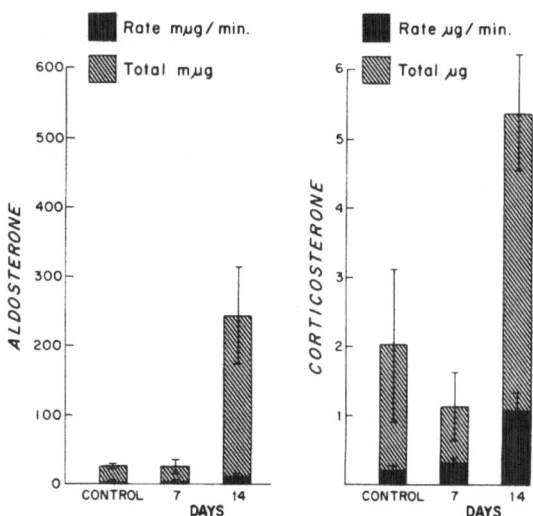

Fig. 1. Effect of treatment with angiotensin amide in oil on adrenal steroid content (striped) and secretion (solid) in intact rats.

2. LABORATORY REAGENTS

Complex distillation and purification of certain solvents has been reduced. It has proven to be more convenient to purchase "analytical reagent" or other "reagent grade" solvents and chemicals than to purify the lower-grade materials. The best grades of pyridine and acetic anhydride do not usually require redistillation. Methylene chloride may be used without silica gel filtration, providing that there is no visible color or residue when the solvent alone is dried. Some lots of ethanol and methanol develop spurious tritium counts due to luminescent impurities. This phenomenon is augmented by storage of the solvent in polyethylene containers. Solvent blanks should be counted in the final elution stage and, if high, acetone is satisfactory as an alternate solvent for corticosteroids although the yield is reduced compared to when the alcohols are used.

The previous method for redistillation of the glacial acetic acid, for use in the oxidation reagent, has proved to be dangerous and yielded a variable product. A simplified procedure now yields a reagent which is reliable and available in quantity. Reagent grade glacial acetic acid which "passes dichromate test" is obtained, and a stock solution is made by the addition of 12 ml of distilled water to 1 liter of glacial acetic acid. The oxidation reagent is prepared fresh from 50 mg of chromium trioxide in 10 ml of the stock solution. The amber reagent is stored at 4°C and discarded after 2 weeks, or earlier if it darkens to a brown color. Oxidation time for a particular steroid is determined empirically by reacting steroid ultraviolet standards for various time intervals, usually 2 to 5 min, at room temperature.

3. RAPID METHOD FOR DETERMINING ACETIC ANHYDRIDE-H³ SPECIFIC ACTIVITY

The standard procedures for assaying specific activity of the acetylating reagent involve chromatography of an acetylated steroid derivative followed by ultraviolet absorption or fluorescence assay of the product. This may require several days and delay the use of a new supply of reagent or computation of data. A simplified method is as follows: 50 mg of pure cortisol is weighed with a microbalance and dissolved in acetone in a 50-ml volumetric flask. 0.5 ml of the solution and 0.5 ml of acetone alone are each dried and reacted with an exact volume of tritium reagent (20 μl of 20% in benzene) and pyridine. After 24 hr incubation at 37°C, the samples are taken to dryness without prior extraction. Each residue is transferred with multiple washes of acetone to a 500-ml volumetric flask and made to volume. Duplicate 1-ml aliquots of the solutions are dried and counted. The acetone—reagent blank is subtracted from the cortisol value. Since the sample represents 1 μg of original steroid, the data is immediately obtained as cpm per microgram of cortisol converted to the monoacetate. This rapid method has provided values which are within ± 5% of the specific activity as determined by more time-consuming methods.

4. THIN-LAYER CHROMATOGRAPHY IN THE ANALYSIS OF CORTISOL AND 11-DESOXYCORTISOL IN HUMAN PLASMA

The analysis of plasma cortisol alone can be done readily with paper chromatography and oxidation of cortisol acetate to cortisone acetate. With this procedure, plasma cortisol is measurable over a linear range of 1 to 100 μg/100 ml in 2 ml aliquots of plasma, and "blank" plasma from dexamethasone-treated subjects assays less than 1 μg/100 ml. This method has been useful in the assay of cortisol in children, in subjects receiving drugs which interfere with chemical tests, and in patients receiving dexamethasone-suppression tests for suspected Cushing's syndrome [5]. A further application of this analysis is made possible by use of thin-layer chromatography. During tests of ACTH reserve, the drug SU-4885 is administered to block 11-hydroxylation of 11-desoxycortisol to cortisol. The subsequent release of ACTH is measured by the increment in urinary 17-hydroxysteroids. The test requires 3 days of urine collection and is subject to variations due to incomplete action of the drug and to the mixed assay of two groups of steroid metabolites, unless the urine is fractionated. The measurement of 11-desoxycortisol in plasma was hindered by difficulty in purification by paper chromatography alone, since a secondary derivative is not formed on exposure to chromic trioxide. The use of two-dimensional thin-layer chromatography as an intermediate step was successful since control plasma samples assayed 0.0 to 0.5 μg/100 ml in 40 subjects [6]. The flow diagram for the simultaneous assay of cortisol and 11-desoxycortisol is shown in Fig. 2. The chromatography systems for cortisol ("F") are as follows:

1. Cyclohexane:100, benzene:40, methanol:100, water:20—16 hr.
2. Cyclohexane:100, dioxane:80, methanol:50, water:20—16 hr.
3. System 1 after oxidation to cortisone acetate—16 hr.

The systems for 11-desoxycortisol ("S"):

1. Same as for cortisol, in same sample.
2. Thin-layer silica gel-G 400 μ thickness.
   a. 5% methanol in chloroform.
   b. hexane:1, ethyl acetate:3.
3. Cyclohexane:100, dioxane:60, methanol:50, water:15—16 hr.

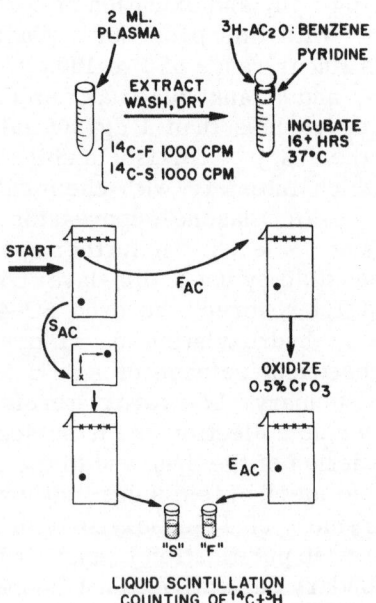

Fig. 2. Simultaneous cortisol and II-desoxycortisol analysis.

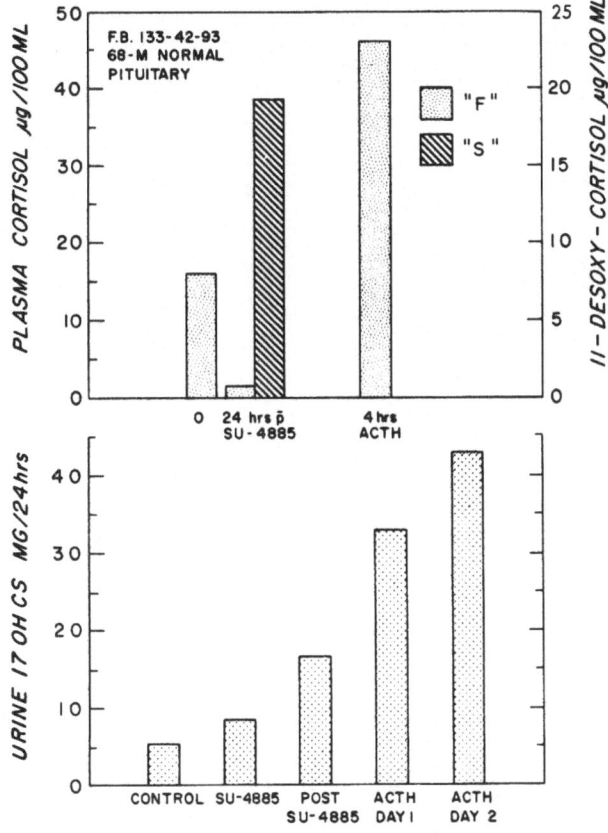

Fig. 3

The same thin-layer system may be substituted in the cortisol analysis for paper system 2. The chromatography then requires only 48 hr for completion. Since extraction and acetylation require only 1 day, the entire assay can be done in 3 working days. Figure 3 shows results of simultaneous assays of the 2 steroids in a normal subject receiving SU-4885. An unresponsive test for a patient with a pituitary tumor is shown in Fig. 4. If only 11-desoxycortisol ("S") is measured, less information is obtained, but a single result of 10 μg/100 ml or above is equivalent to a two- to threefold rise of urinary steroids on the third day (Fig. 5). Since the turnover rate of 11-desoxycortisol ("S") is about twice that of cortisol ("F"), the

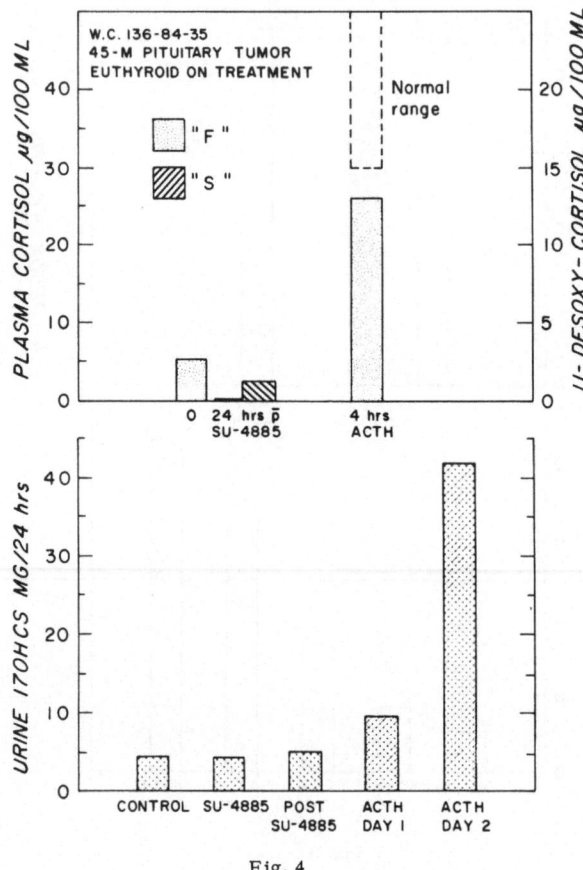

Fig. 4

plasma level of "S" is about one half that of "F" at similar
secretion rates, and comparison of changes is based on
F = 2 S as a formula for relative steroid concentration. An
advantage of this analysis is that two plasma samples in 24 hr
provide as much or more information as three 24-hr urine
collections. The inhibitory action of the drug is also observed
when both steroids are assayed in plasma. The thin-layer
chromatography provides rapid processing, and the double
isotope method provides a quantitative measure of steroid
levels. The small amount of plasma used has allowed the use
of this test in children with results similar to those obtained
in adults.

5. GAS–LIQUID CHROMATOGRAPHY IN THE ASSAY OF
TESTOSTERONE IN HUMAN PLASMA

The assay of testosterone in human plasma requires a
sensitive technique since levels are less than $1 \mu g/100$ ml in
adult males and below $0.1 \mu g/100$ ml in females. Several
methods have been developed using double-isotope derivatives
and secondary derivative formation. The method used in our
laboratory has evolved from studies with gas chromatography
of radioactive steroids. Studies with pure testosterone demonstrated that the C^{14}-acetate derivative could be either
fractionated by gas–liquid chromatography, then pyrolyzed and
detected as C^{14}-CO_2, or condensed on solid scintillation crystals
without pyrolysis [8]. Attempts to purify biological samples
without secondary derivatives failed to achieve reliable purification. The gas-chromatographic collection technique eliminated tritium contaminants which passed through the paper and
thin-layer systems. The schematic method is shown in
Fig. 6. The first chromatography is carried out for 6 hr in

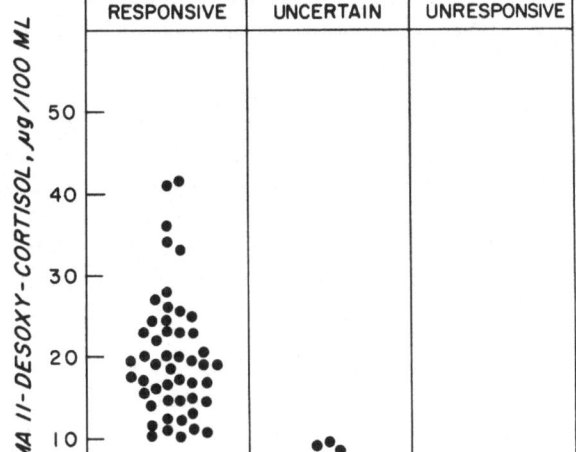

Fig. 5

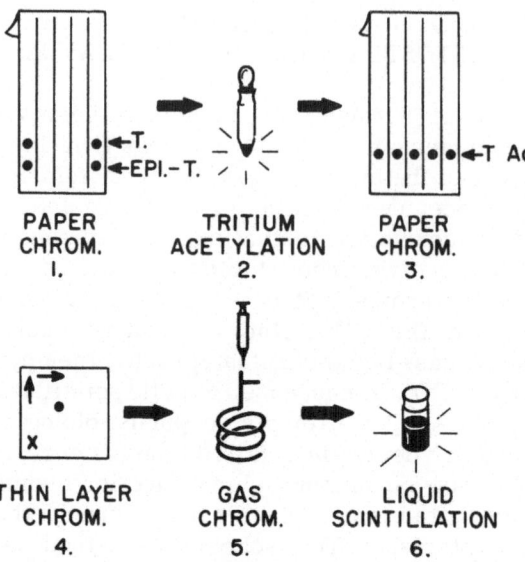

PAPER   TRITIUM   PAPER
CHROM.   ACETYLATION   CHROM.
I.   2.   3.

THIN LAYER   GAS   LIQUID
CHROM.   CHROM.   SCINTILLATION
4.   5.   6.

Fig. 6. Plasma testosterone analysis by double–isotope dilution: 4–C^{14}–tracer, 10 to 20 ml plasma.

the descending paper system–cyclohexane:4, benzene:2, methanol:4, water:1.   This step removes acetylable substances and separates the testosterone from epi–testosterone.   After tritium acetylation, the second chromatography on paper is in cyclohexane:2, nitromethane:1, methanol:1 for 5 hr.   A two–dimensional thin–layer chromatography is run with the ascending solvent systems hexane:1, ethyl acetate:1, and benzene:2, ethyl ether:1.   The final eluate is dried, dissolved in ethyl acetate, injected into a Glowall gas chromatograph, and is fractionated on a 6–ft, 2% SE–30 column.   The testosterone acetate is collected onto uncoated terphenyl crystals, using a Packard gas fraction collector, and the crystals are ejected into toluene phosphor for liquid–scintillation counting.   The gas chromatography requires only 5 to 10 min for each sample. The entire assay may be performed in 4 days or less after sample extraction.   Further details of this method are reported elsewhere [9].   The application of this method to the diagnosis of disorders of male hormone secretion is shown in Fig. 7. The mean values are indicated by horizontal lines, and one standard deviation is shown by the shaded areas.   The values for 30 men with hypogonadism due to pituitary hypofunction are reduced compared to values for 22 normal adult males. A group of patients with Klinefelter's syndrome due to an XXY sex chromosome defect had partial signs of virilization and an intermediate range of values which were below normal. The method directly measures the degree of testosterone deficiency in men or an excess of testosterone in women.

6. COMPUTER PROGRAM

The computation of steroid values from isotope data can be time-consuming and subject to calculation errors. A digital computer program provides high-speed transformation of data with increased reliability. A teletype printout for a PDP-4 computer is seen in Fig. 8. The first section shows the data input typed by the operator who enters the values for H^3 and C^{14} dpm standards, reagent specific activity, and cross-channel ratio for tritium. The experimental data from the scintillation counter printout is entered in sequence of time, channel A counts, and channel B counts. The aliquot value for each sample is typed, the tracer H^3/C^{14} ratio is given, and the

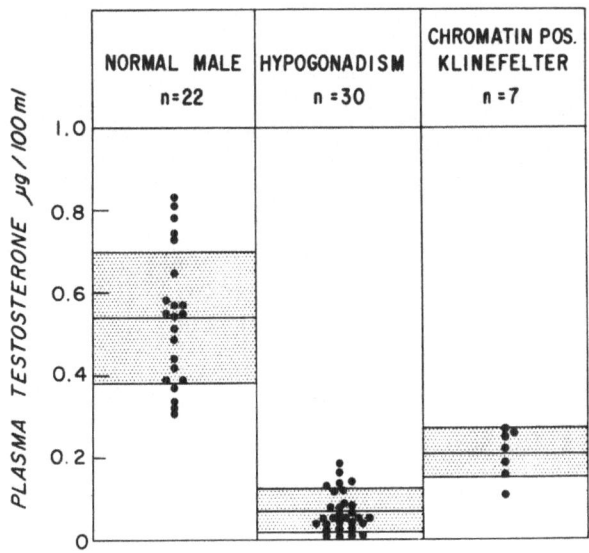

Fig. 7

DIGITAL COMPUTER PROGRAM FOR DOUBLE ISOTOPE DERIVATIVE ANALYSIS OF STEROIDS

```
32100. 83200. 423500. 0.001      14C DPM STD ; 3H DPM STD ; REAGENT S.A.; 3H A/B

  1   3000 441681. 155669.        DATA INPUT
  2   3000 3017. 605142.
  3   3000 592. 1216.
  4   3000 39656. 21177.
  5   3000 39420. 20402.
  6   3000 611. 2292.
  7   3000 559. 2014.
  8   3000 3840. 14192.
  9   3000 3565. 25491.
 10   3000 3006. 14554.
 11   3000 3320. 17035.
 12   3000 4224. 34402.
 13   3000 1938. 11779.
 14   3000 2590. 10707.
 15   3000 2961. 54314.
 16   3000 7051. 14635.
 17   -1                          END DATA INPUT
1.15                             TRACER B/A
8 100 25                         SAMPLE VOLUME
9 100 25
10 100 25
11 100 20
12 100 24
13 100 14
14 100 25
15 100 25
16 100 25
```

Fig. 8a. Digital computer program for double isotope derivative analysis of steroids.

COMPUTER OUTPUT

		ACPM	BCPM	ACPM-B	BCPM-B		
1	0	.14722 05	.51889 04	.14702 05	.51484 04	CALC STDS	
2	0	.10056 03	.20171 05	.80833 02	.20130 05		
3	0	.19733 02	.40533 02	.00000 00	.00000 00		
4	0	.13218 04	.70589 03	.13014 04	.62949 03		
5	0	.13140 04	.68006 03	.12936 04	.60366 03		
6	0	.20366 02	.76399 02	.00000 00	.00000 00		
7	0	.18633 02	.67133 02			CALC TRACER	

C14EF	C14RAT	H3EF	RSA	C14IS	
.45803-00	.35016-00	.24195-00	.10246 06	.12975 04	CALC REAGENT

		ACPM	BCPM	ACPM-B	BCPM-B	H3YLD	ALIQT
8	0	.12800 03	.47306 03	.10936 03	.40593 03	.28179-01	.11271-00
9	0	.11883 03	.84969 03	.10019 03	.78256 03	.80680-01	.32272-00
10	0	.10019 03	.48513 03	.81566 02	.41799 03	.46231-01	.18492-00
11	0	.11056 03	.56783 03	.92033 02	.50069 03	.50272-01	.25136-00
12	0	.14079 03	.11467 04	.12216 03	.10795 04	.93905-01	.38501-00
13	0	.64599 02	.39263 03	.45966 02	.32549 03	.71312-01	.50631 00
14	0	.86333 02	.35689 03	.67699 02	.28976 03	.35436-01	.14174-00
15	0	.93699 02	.13104 04	.80066 02	.17433 04	.26285-00	.10514 01
16	0	.23503 03	.48733 03	.21639 03	.42069 03	.56693-02	.22677-01

Fig. 8b. Computer output.

computer proceeds with computation. Twelve lines of data are printed each minute and appear in the second section. Counting efficiencies and cpm for tracer and blood samples are given, other intermediate data is provided as requested, and the values of steroid quantity and concentration are printed in the final two columns. For large numbers of assay samples, punch tape or punch card input may be used in preference to teletype input. Additional programs may be added to calculate statistical analysis of the data.

REFERENCES

1. Kliman, B., and Peterson, R.E., J. Biol. Chem. 235:1639, 1960.
2. Peterson, R.E., in Rothchild, S. (editor), Advances in Tracer Methodology Vol. 1, Plenum Press, New York, 1963.
3. Kliman, B., in Rothchild, S. (editor), Advances in Tracer Methodology Vol. 2, Plenum Press, New York, 1965.
4. Dufau, M.L., and Kliman, B., VI Pan-American Congress of Endocrinology, Mexico City, October 1965, Excerpta Medica International Congress Series No. 99.
5. Kliman, B., Scoggins, R.B., Apple, B.A., and Rausch-Stroomann, J.G., Plasma Double-Isotope Cortisol for Evaluation of Adrenal Suppression, Program of 46th Endocrine Society Meeting, San Francisco, 1964.
6. Kliman, B. Hormonal Steroids, Milan, Italy, May 1966, Excerpta Medica International Congress Series No. 111.
7. Plager, J.E., Schmidt, K.G., and Staubitz, W.J., J. Clin. Endocrinol. Metab. 25:499 (1965).
8. Karmen, A., McCaffrey, I., and Kliman, B., Anal. Biochem. 6:31 (1963).
9. Kliman, B., and Briefer, Jr., C., in Grant, J.N. (editor), International Symposium on the Gas—Liquid Chromatography of Steroids, Glasgow, April 1966, Cambridge University Press, Cambridge, England.
10. Briefer, Jr., C., Forbes, A.P., and Kliman, B., Program of 47th Endocrine Society Meeting, New York, 1965.

ESTIMATION OF SPECIFIC RADIOACTIVITY
OF STEROIDS BIOSYNTHESIZED FROM LABELED
PRECURSORS

Kristen B. Eik-Nes

Department of Biological Chemistry
University of Utah College of Medicine
Salt Lake City, Utah

In 1963, it was suggested that steriods containing electrophoric groups could be detected by the electron capture cell following gas-phase chromatography [1, 2]. Subsequent to these publications, our laboratory observed that the chloroacetate of pure testosterone and of pure 20-reduced progesterone could be detected in submicrogram amounts by electron capture [3]. Methods were later developed for the estimation of testosterone [3], estradiol [4, 5], and progesterone [6] in biological fluids utilizing electron capture of steroid chloroacetates.

Clark and Wotiz [7] reported that heptafluorobutyrates of pure steroids could be detected in very low concentrations by electron capture, but no formal publication on the use of such derivatives for the estimation of steroids in biological fluids has hitherto appeared. Rapp and Eik-Nes [8] reported that some derivatives of pure corticosteroids could be determined by electron capture after gas-phase chromatography. Prominent among these derivatives were deoxycorticosterone acetate, the γ-lactone of aldosterone, the acetate of 11-dehydrocorticosterone, and adrenosterone. The same authors used these observations in a method for the quantitation of deoxycorticosterone and aldosterone in biological samples [9]. Moreover, during the last year we have observed that the pentafluorophenylhydrazone of steroid ketones will capture electrons, and a method has been developed for the estimation

This work was supported in part by research grants AM-06651 and TO1 CA 5000 from the U.S. Public Health Service, Bethesda, Maryland.

All isotopes employed in the experiments to be reported were purchased from the New England Nuclear Corp. and purified before use.

241

of estrone in blood plasma [10]. This method can detect as little as 200 pg of estrone pentafluorophenylhydrazone.

Figures 1 to 5 illustrate the type of sensitivity for endpoint analysis these methods have. For such sensitivity, however, the steroid to be estimated must be in a very high state of purity prior to gas-phase chromatography. This is achieved through solvent partition, paper, and thin-layer chromatography. In our work with electron capture detection of steroids following gas-phase chromatography, we have found that the short columns containing a nonselective phase are the best ones. Little compound separation is done on such columns [11]. Thus, meaningful data by this technique can only be obtained from biological samples of extreme purity [12, 13] (Table I). Moreover, these techniques [3, 4, 9, 10] call for internal standardization to determine the sensitivity of the electron capture cell as well as to determine sample transfer to the gas chromatograph. These factors must be known for each sample subjected to gas-phase chromatography.

With the current available electron capture cell, the best precision of assay is obtained when between 4 to 8 ng of material are chromatographed [3]; sample size smaller than 2 ng gives considerable inaccuracy. When quantitation is done at these levels on samples from biological material, one should

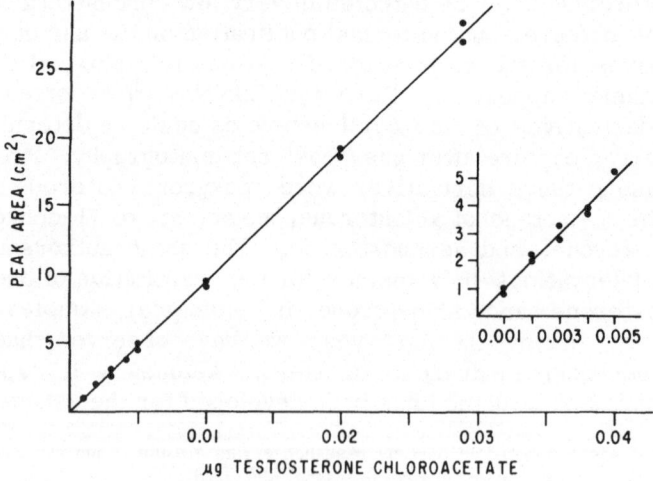

Fig. 1. Calibration curve (cm² peak vs. μg) of pure testosterone chloroacetate [3].

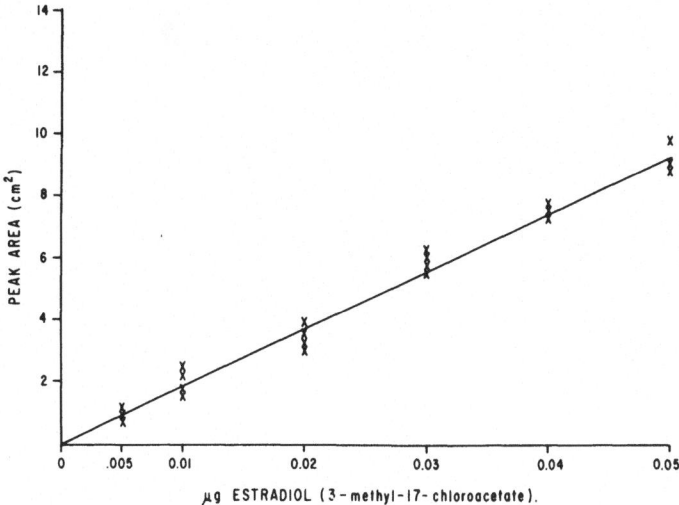

Fig. 2. Calibration curve (cm² peak vs. μg) of pure estradiol-3-methylether-17-chloroacetate [4].

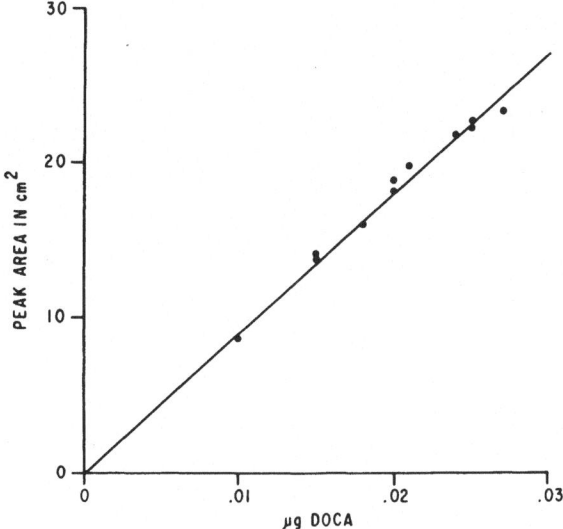

Fig. 3. Calibration curve (cm² peak vs. μg) of pure deoxycorticosterone acetate [9].

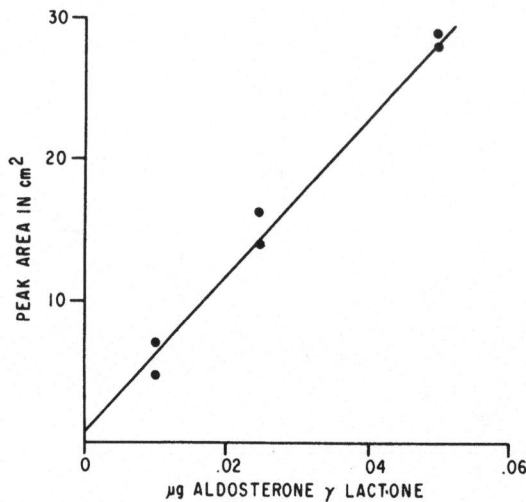

Fig. 4. Calibration curve (cm² peak vs. μg) of γ-lactone of aldosterone [9].

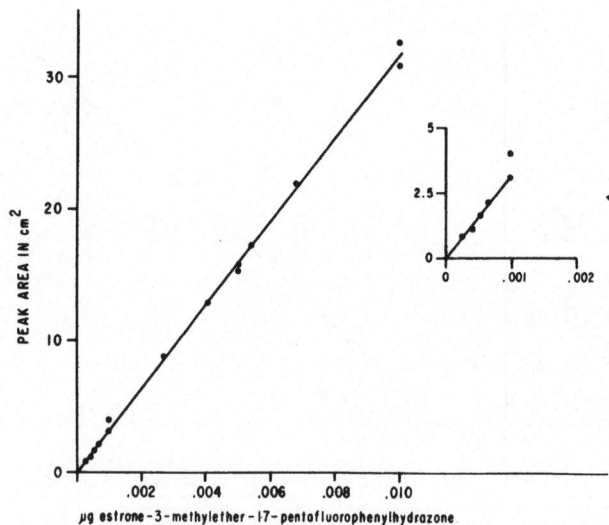

Fig. 5. Calibration curve (cm² peak vs. μg) of pure estrone-3-methylether-17-
pentafluorophenylhydrazone [10].

keep in mind that the electron capture cell may easily become poisoned from impurities still present in the biological sample [14].

Utilizing isotopes of different energy spectra, methods have been developed for submicrogram quantitation of steroids in biological fluids [15—17]. These techniques may well be more adequate than electron capture for steroid estimation at low concentrations, but they are difficult to apply to quantitation of steroids biosynthesized from labeled precursors [18]. Since electron capture detection does not involve the use of isotopes, this technique is still valid when the steroid to be estimated is formed from either a C^{14}- or an H^3-containing precursor. Much work in steroid endocrinology deals with the establishment of pathways used for steroid formation. In such work endocrine tissues are exposed to labeled precursors of different energy spectra in vivo or in vitro and the content of isotopes is measured in steroid intermediates and in hormonal end products. Frequently, little attention is paid to the fact that the tissues under investigation are producing their own unlabeled intermediates and end products from unlabeled precursors. Moreover, there is a good deal of evidence to suggest that the unlabeled precursors, contained in the biosynthetic cell, are handled differently from the labeled pre-

Table I. Crystallization of Testosterone-C^{14} (Chloroacetate) to Constant Specific Activity from Several Solvent Combinations

Solvent	Specific radioactivity dpm/mg
Benzene/hexane	10,780
Ethylacetate/hexane	11,000
Aqueous methanol	10,900

*Acetate-1-C^{14} was infused via the spermatic artery of a dog and venous blood collected from the infused tissue. The plasma sample was processed by the testosterone method of Brownie et al. [3], but after the last thin-layer chromatogram, 10 mg authentic testosterone chloroacetate (corrected mp 124.3°C) was added to the purified plasma testosterone sample, and the mixture was crystallized from several solvent combinations.

cursors added to the medium surrounding the cell. Finally, in such isotope experiments, the ability of the added precursors (H^3 and C^{14} labeled) to penetrate the cell is not measured, and the pool size of precursors (labeled and unlabeled) is often ignored.

Quantitation of steroids at the submicrogram level by electron capture offers distinct advantages in endocrine investiga-

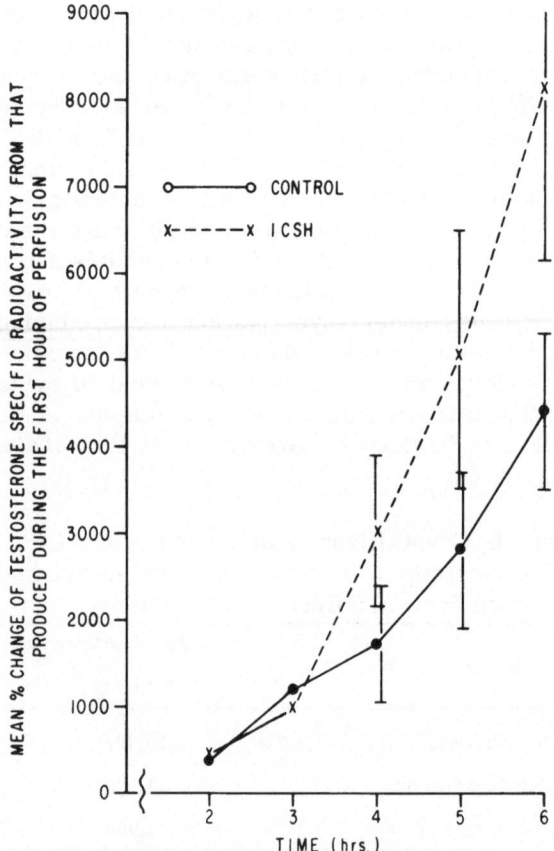

Fig. 6. Effect of interstitial cell stimulating hormone (ICSH) on mean specific radioactivity of testosterone-C^{14} produced from acetate-1-C^{14} by rabbit testis perfused with acetate-1-C^{14} and cholesterol-7α-H^3 at the same time. The data are from four control testes and eight testes also perfused with ICSH. When used, ICSH was perfused at a constant rate from the third to the sixth hour of experiment. One standard error of the mean value is indicated with some points. The data are from a publication by Ewing and Eik-Nes [12].

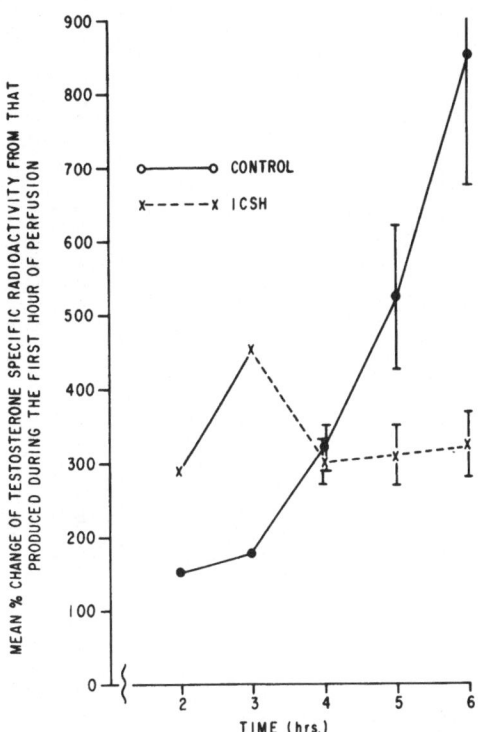

Fig. 7. The effect of ICSH on mean specific radioactivity of testosterone-H^3 produced from cholesterol-7a-H^3 by rabbit testes perfused with acetate-1-C^{14} and cholesterol-7a-H^3 at the same time. The data are from four control testes and eight testes perfused also with ICSH. When used, ICSH was perfused at a constant rate from the third to the sixth hour of experiment. One standard error of the mean value is indicated with some points. The data are from a publication by Ewing and Eik-Nes [12].

tions employing isotopes to trace metabolic pathways. An example of this application is given in Figs. 6 and 7. When acetate-1-C^{14} and cholesterol-7a-H^3 were perfused via the spermatic artery, a gonadotrophin increased the specific radioactivity of testosterone-C^{14} and decreased the specific radioactivity of testosterone-H^3 in spermatic venous blood. This could indicate that there is some effect of the gonadotrophin on the formation of cholesterol from acetate or also that the cholesterol-7a-H^3 infused in our experiment does not saturate the sterol pool(s) in the testis used for testosterone formation under trophic influence. By measuring only the content of H^3 and C^{14} in testosterone in this study, we would be

Table II. The Effect of Increasing Amounts of 17α-Hydroxyprogesterone-4-C^{14} on Formation of Androstenedione and Testosterone from Dehydroepiandrosterone-7α-H^3 by Ovarian Tissue Homogenates

Flask number	17α-Hydroxy-progesterone added (μg)	Androstenedione Specific radioactivity*		R^b	μg isolated	Testosterone Specific radioactivity†		R^b	μg isolated
		H^3	C^{14}			H^3	C^{14}		
1	0	598,300			1.72	466,400			0.020
2	4	602,500	9200	65.4	2.03	382,900	4800	79.8	0.027
3	8	531,900	12,700	41.9	1.91	536,800	10,900	49.2	0.024
4	16	573,200	16,800	34.1	1.67	560,000	17,700	31.6	0.020

*The data are given in disintegrations/per minute per micrograms.
†Ratio H^3/C^{14} in isolated steroid. All flasks contained the same amount of dehydroepiandrosterone-7α-H^3 (15 μg and 5.0 μC). 17α-Hydroxyprogesterone-4-C^{14} was added in the following amounts: 0, 4, 8, and 16 μg and 0, 0.17, 0.34, and 0.68 μC in flasks 1, 2, 3, and 4, respectively. Ovarian tissue was obtained from dogs treated with pregnant mare serum gonadotrophin. In each flask the equivalent of 100 mg tissue was used. Cofactors were glucose-6-phosphate and TPN, 0.6 mg of each, and tracer amounts of glucose-6-phosphate-dehydrogenase. Total incubation volume was 10 ml and incubation time was 90 min. The data represent the mean of quadruplicate incubations [21].

apt to conclude that the gonadotrophin promoted both acetate-1-C^{14} and cholesterol-7a-H^3 conversion to testosterone [13] since the gonadotrophin increased both isotopes in testosterone. The amounts of testosterone produced in these experiments were too low to be measured by conventional techniques, and their estimation by the principle of double isotope dilution would have been difficult to carry out [18].

For some years it has been known that testosterone can be formed in the ovary and in the testis from dehydroepiandrosterone via Δ^5-androstenediol [19, 20]. Table II presents some evidence that this formation may be influenced by 17a-hydroxyprogesterone. In this study 17a-hydroxyprogesterone-4-C^{14} and dehydroepiandrosterone-7a-H^3 were incubated with homogenate of dog ovaries and the specific radioactivity of androstenedione-C^{14} and H^3 and of testosterone-C^{14} and H^3 were measured [21]. Addition of 17a-hydroxyprogesterone had little effect on specific radioactivity for H^3 of androstenedione, but appeared to increase this activity in testosterone. If all of the testosterone-H^3 formed in this sytem used androstenedione-H^3 as an obligatory intermediate, one would expect that the specific radioactivity of androstenedione and testosterone with respect to H^3 should be the same and should change in a parallel fashion upon the addition of increasing concentrations of 17a-hydroxyprogesterone. The data of Table II strongly support an inhibitory effect in vitro of 17a-hydroxyprogesterone on the formation of testosterone from dehydroepiandrosterone via Δ^5-andostenediol. Moreover, radioactive 17a-hydroxyprogesterone appears to compete for a saturated enzyme. These effects of 17a-hydroxyprogesterone on androgen formation in vitro could not have been discovered unless specific radioactivity of both androgens were measured.

It is well established that estrogens are formed from androgens in endocrine organs. Moreover, recent data indicate that the drug clomiphene [1β(β-diethylaminoethoxy) phenyl-1,2-diphenyl-2-chloroethylene] will stimulate estrogen excretion in the woman though no great agreement exists as to the mechanism of action of clomiphene on estrogen formation—whether via the hypothalamic-pituitary system or via direct ovarian stimulation. In order to elucidate this problem, androstenedione-4-C^{14} was infused via the ovarian artery of dogs in vivo and specific radioactivity of estradiol-C^{14} was measured in ovarian venous blood. If large amounts of clomi-

Table III. Results of Analysis for Mass and Specific Radioactivity of Aldosterone, Deoxycorticosterone, and Corticosterone from Rat Adrenals Incubated with Progesterone-4-C^{14}

Compound isolated	Sodium-deficient		Control		Sodium Deficient	Angiotensin
	Saline	Angiotensin	Saline	Angiotensin		
Aldosterone, µg/incubate	9.01†	7.88	4.11	3.42	$P < 0.01$	NS†
Corticosterone, µg/incubate	5.9(2)	5.3(2)	18.1(3)	16.9(4)	$P < 0.01$	NS
DOC, µg/incubate	0.35	0.32	1.77	1.67	$P < 0.01$	NS
Aldosterone,* sp. act.	20,260	22,340	20,820	23,660	NS	NS
Corticosterone,* sp. act.	20,300(2)	25,350(2)	18,000(3)	16,780(4)	NS	NS
DOC,* sp. act.	36,160	38,400	25,280	19,860	$P < 0.01$	NS

*Specific activities (sp. act.) were expressed as dpm/mµmole. Sp. act. of precursor progesterone-4-C^{14} was 102,200 dpm/mµmole.

†Table entries are the mean of five replicates unless followed by a number in parentheses which indicates the number of replicates.

‡NS indicates not significant, that is, $P > 0.1$. Four halved rat adrenals were used per 25 ml flask; Krebs–Ringer bicarbonate medium (2 ml/100 mg tissue) contining glucose (2 mg/ml) and progesterone-4-C^{14} (769,200 dpm/ml, 2.4 µ;g/ml) was used. Incubation was for 4 hr at 37° C in 95% O_2, 5% CO_2. Angiotensin II was added at the level 3 µg/ml of medium. Preincubation was not done and ACTH was not added. Average weight of adrenals per flask was 104.9 mg for Na-deficient and 107.6 mg for control groups; these values were not different (t test $P < 0.2$). Aldosterone and deoxycorticosterone were measured by electron capture and corticosterone was measured by absorbancy at 240 mµ [9].

phene were infused via the ovarian artery concomitant with infusion of androstenedione-4-C^{14}, the total production as well as the specific radioactivity of estradiol in ovarian venous blood increased [22]. Also, the specific radioactivity of estradiol-C^{14} in the infused tissue was higher in clomiphene treated than in nonclomiphene treated ovaries. These observations give credence to the claim that one of the actions of clomiphene is on the conversion of androgens to estrogens in the ovary.

When steroid specific radioactivity was determined in incubations of rat adrenals with progesterone-4-C^{14} (Table III), it became apparent that the deoxycorticosterone pool in adrenals from sodium-deficient animals is different from that of either aldosterone or corticosterone. The relatively high specific radioactivity of deoxycorticosterone produced by adrenal tissue from sodium-deficient rats is compatible with a theory that biosynthetic blockage or interference with zonal, adrenal storage of needed precursors for deoxycorticosterone formation may be associated with feeding rats a sodium-deficient diet for 13 days. Again, examination of radioactivity alone in these experiments would not have revealed this rather unexpected finding which indeed warrants further experiments.

For the past 7 years our laboratory has been interested in pathways used for the formation of testosterone from Δ^5-pregnenolone in the testis. Two major pathways for this biotransformation are possible—one via progesterone and 17a-hydroxyprogesterone and another via 17a-hydroxypregnenolone and dehydroepiandrosterone. Moreover, we have observed that the testis in vivo can convert 17a-hydroxypregnenolone to 17a-hydroxyprogesterone [23]. Thus, testosterone formation is still possible via two pathways from Δ^5-pregnenolone in which the formation of progesterone is bypassed [24]. However, examination of specific radioactivity of progesterone and 17a-hydroxyprogesterone in spermatic venous blood of dogs infused with Δ^5-pregnenolone-4-C^{14} via the spermatic artery (Table IV) could indicate that the testicular pools of progesterone and 17a-hydroxyprogesterone are different. If all of the 17a-hydroxyprogesterone in these infusion experiments came from progesterone only, one would expect that their specific radioactivities were similar. Since this activity is much lower in 17a-hydroxyprogesterone than in progesterone, nonlabeled 17a-hydroxyprogesterone must dilute the specific radioactivity of the 17a-hydroxyprogesterone formed from

Table IV. Specific Radioactivity of Progesterone and 17α-Hydroxyprogesterone (μc/mmole) in Spermatic Venous Blood of Dogs Infused with Δ^5-Pregnenolone-4-C^{14} via the Spermatic Artery

Steroid	Dog Index	Specific radioactivity in time sample (min)*		
		0−30	30−60	60−90
Progesterone	A	6,270	15,040	12,910
17α-hydroxyprogesterone	A	1,940	1,920	2,536
Progesterone	B	10,620	18,500	13,540
17α-hydroxyprogesterone	B	1,076	3,400	2,180

*During time 0—90 min, the animals were infused at a constant rate with 150 ng and 0.22 μC of Δ^5-pregnenolone-4-C^{14}/min via the spermatic artery and 30-min samples of spermatic venous blood collected. The constant rate of flow of arterial blood to the testis was 3.87 ml/min in each animal, and the infused testes were maintained at 37.5°C [25].

progesterone. This would occur if a large pool of unlabeled 17α-hydroxyprogesterone exists in the infused testis or also if 17α-hydroxypregnenolone of lower specific radioactivity than that of progesterone is converted to 17α-hydroxyprogesterone. Since progesterone has an inhibitory effect on the formation of spermatozoa, knowledge on rates of progesterone metabolism and pool size of the hormone in the testis is important.

From these few examples it clearly emerges that valuable information can be obtained in isotope experiments if specific radioactivity of biosynthetized compounds is measured in addition to the concentrations of isotopes in these compounds. It is hoped that the use of steroid derivatives with high sensitivity for electron capture [7] will increase the application of the methods discussed in this communication since many steroid intermediates exist in biological samples in extremely low concentrations.

REFERENCES

1. Landowne, R. A., and Lipsky, S. R., Anal. Chem. 35:532 (1963).
2. Lovelock, J. E., Simmonds, P. G., and van den Heuvel, W. J. A., Nature, 197:249 (1963).

3. Brownie, A.C., van der Molen, H.J., Nishizawa, E.E., and Eik-Nes, K.B., J. Clin. Endocrinol. Metab. 24:1091, 1964.
4. Eik-Nes, K.B., Aakvaag, A., and Grota, L.J., in Lipsett, M.L. (editor), Gas Chromatography of Steroids in Biological Fluids, Plenum Press, New York, 1965.
5. Aakvaag, A., and Eik-Nes, K.B., Biochim. Biophys. Acta 111:273 (1965).
6. van der Molen, H.J., and Groen, D., in Lipsett, M.L. (editor), Gas Chromatography of Steroids in Biological Fluids, Plenum Press, New York, 1965.
7. Clark, S.J., and Wotiz, H.H., Steroids, 2:535 (1963).
8. Rapp, J.P., and Eik-Nes, K.B., J. Gas Chromatog. 3:235 (1965).
9. Rapp, J.P., and Eik-Nes, K.B., Anal. Biochem. 15:386 (1966).
10. Attal, J., Hendeles, S.M., and Eik-Nes, K.B., Anal. Biochem. 20:394 (1967).
11. Eik-Nes, K.B., and Horning, E.C., Gas-Phase Chromatography of Steroids in Biological Fluids. Springer-Verlag, Heidelberg and New York, 1967.
12. Ewing, L.L., and Eik-Nes, K.B., Can. J. Biochem. 44:1327 (1966).
13. Connell, G.M., and Eik-Nes, K.B., Proc. Natl. Acad. Sci. 55:410 (1966).
14. Rapp, J.P., and Eik-Nes, K.B., J. Gas Chromatog. 4:376 (1966).
15. Riondel, A., Tait, J.F., Gut, M., Tait, S.A.S., Joachim, E., and Little, B., J. Clin. Endocrinol. Metab. 23:620 (1963).
16. Hudson, B., Coghlan, J., Dulmanis, A., Wintour, M., and Ekkel, J., Austral. J., Exp. Biol. Med. Sci. 41:235 (1963).
17. Horton, R., J. Clin. Endocrinol. Metab. 25:1237 (1965).
18. Tait, J.F., personal communication, October 1966.
19. Gospodarowicz, D., Acta Endocrinol. 47:306 (1964).
20. Hagen, A.A., and Eik-Nes, K.B., Biochim. Biophys. Acta 90:593 (1964).
21. Aakvaag, A., and Eik-Nes, K.B., Biochim. Biophys. Acta 111:286 (1965).
22. Engels, J.A., Friedlander, R.L., and Eik-Nes, K.B., in press (1967).
23. Hagen, A.A., and Eik-Nes, K.B., Biochim. Biophys. Acta 86:372 (1964).
24. Eik-Nes, K.B., Physiol. Rev. 44:609 (1964).
25. van der Molen, H.J., and Eik-Nes, K.B., Excerpta Med. Found. Intern. Congr. Series 111, No. 484, 1966.

CONTINUOUS RADIOACTIVITY MONITORING OF PERFUSION IN THE SMALL INTESTINE OF THE INTACT ANIMAL

Francis A. Jacobs

Guy and Bertha Ireland Research Laboratory
Department of Biochemistry
University of North Dakota School of Medicine
Grand Forks, North Dakota

The phenomenon of intestinal absorption and transport has been studied by a variety of techniques. Since in vitro studies are removed from the physiological state and since it is very difficult, if not impossible, to simulate the steady state in the living intact animal, various modifications of in vivo systems have been developed. We have reported [1] an in vivo perfusion system in which continuous absorption studies can be carried out using a single anesthetized intact small animal (rat), permitting the determination of continuous absorption patterns from the intestinal lumen in the same living animal. Sampling is possible at any elected time interval. The system described is a recirculating one. We have described a modification of the system, a single-pass perfusion [2, 3] of the lumen with no recirculation of the solution perfused. Both of the systems have certain limitations in their operation. In the recirculating system samples drawn from the mixed perfusate during the experimental period bring about a diminution of the total volume perfused. In the single-pass system samples can be collected over any selected time interval; however, the quantities of material which are required for analyses and the slight gradient in concentration can pose technical problems.

The type of data obtainable from the first-described system allow one to study absorption rates from the intestinal lumen.

This investigation was supported in part by the U.S. Public Health Service Research Grant AM-02023-NTN from the National Institute of Arthritis and Metabolic Diseases.

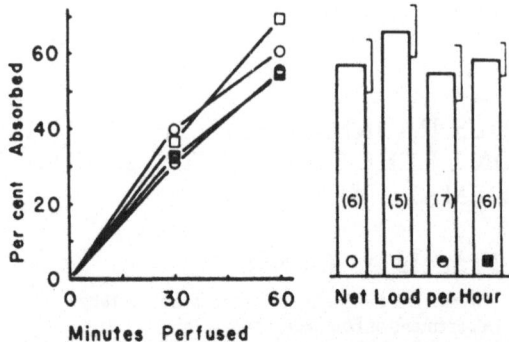

Fig. 1. The effect of pyridoxine, pyridoxal, and pyridoxal phosphate on L-tyrosine absorption. Left: Continuous absorption rates for the control group (O); those injected with pyridoxal (□); those injected with pyridoxine (◐); and those injected with pyridoxal phosphate (■). Right: The net loads absorbed for these respective groups after 1 hr of perfusion (the same symbolism applies). Numbers in parentheses indicate group size for this and subsequent figures. Chemical analyses were made at 0, 30, and 60 min of perfusion. Reproduced with permission from the Journal of Biological Chemistry. [4].

One can see from Fig. 1 that linear rates are obtainable but that there are limitations in the number of samples that can be justified for chemical analyses to establish the pattern or slope of the absorption rate [4]. Chemical analyses of initial and final load perfused allow one to obtain a net rate of absorption over a fixed period of time, as can be seen in the figure. Any change in rate during the interval could not be detected; only the net effect would be apparent.

The system described in this communication is a modification of these systems so designed that both absorption and efflux (secretion across the intestinal mucosa into the lumenal contents) can be measured. Measurements can be made by counting or by chemical determinations at the start and the termination of the perfusion experiment as well as by a continuous scintillation monitoring of the C^{14}-radioactivity levels of the material being perfused. Any change in absorption rate during the perfusion period is detectable and measureable. The system is applicable to kinetic studies.

The monitored perfusion system has been used in an investigation of amino acid absorption and transport. Data from these studies will be used to illustrate the application of the system.

METHODS AND MATERIALS

Apparatus for Monitored Perfusion

A diagramatic representation of the apparatus is shown in Fig. 2. In essence, the perfusion apparatus is very similar to that described earlier [1]. There are two independent circulatory systems. One circulates through a thermostatically controlled heat exchanger to maintain the temperature of the perfusion solution (perfusate). The perfusion system recirculates through a reservoir, pump, filter-flowmeter system, heat exchanger, and the in situ cannulated segment of the upper small intestine. In the modified design the heat exchanger is much closer to the animal than in the earlier system; thus there is improved control of the temperature of the perfusing solution. The reservoir contains a magnetic mixer which permits an equilibration of the effluent perfusate from the cannulated intestinal lumen with the reserve material in the system. The pump is adjustable to allow for a 1.5-ml/min flow rate of the perfusate. Cannulas are made of glass with polished edges so as to avoid damage to the intestinal mucosa. Additives such as antimetabolites or inhibitors can be added without affecting the circulation volumes.

In this system radioactivity monitoring can be performed without the perfused preparation being disturbed by sample collection or volume change. From the flowmeter a shunt circuit can divert the flow of perfusion solution through a scintillation detector. The perfusate passes through an anthracene-packed cell which is connected to the rest of the system through black, capillary Teflon tubing. The scintillation detector*for fluid flow has been described by Rapkin and Gibbs [5] and has been applied by Elwyn [6] as well as ourselves [3, 7] for the monitoring of the effluent from chromatographic columns used for amino acid analysis. The detector system feeds its signal through a scintillation spectrometer and recording ratemeter with automatic printout.*

With this series of electronic components it is possible to measure the level of radioactivity present in the solution being perfused through the intestinal lumen and to plot this activity

*Manufactured by Packard Instrument Company, Series 317 Flow Detector, Series 500 E TriCarb Scintillation Spectrometer, and Model 380 Recording Ratemeter.

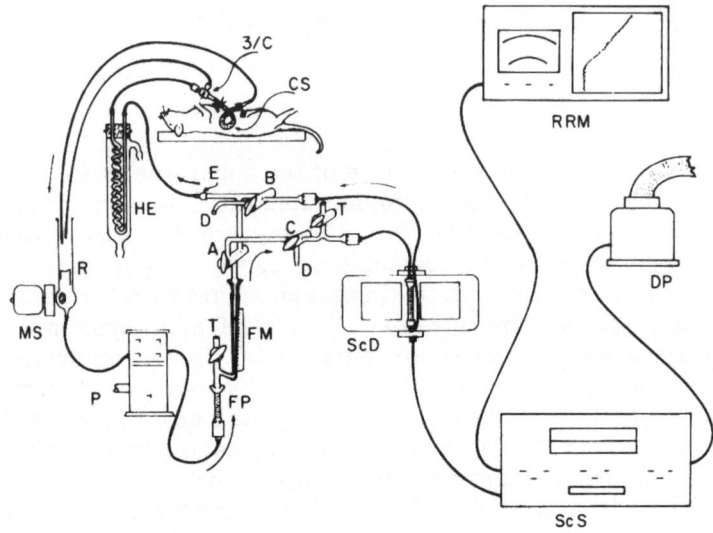

Fig. 2. Monitored Perfusion System. The perfusion lines except those to the detector are made of 1/16 in. ID × 1/8 in. OD polyvinyl (Tygon) tubing. The lines leading to and from the scintillation detector (ScD) are made of 20-gauge ID opaque, black Teflon tubing. The animal table and heat exchanger (HE) are kept warm by a thermostatically controlled circulating water system; the coil is made of a 14-gauge annealed hypodermic stainless steel tubing, 5 ft long. A three-way stopcock (BD 5031 MS, 3/C) is used to shunt the flow of the perfusion solution away from the cannulated segment and back to the reservoir (R) which is mixed mechanically (MS) by a Teflon-covered magnetic stirring bar. The pump (P) used in this system must overcome the back-pressure of the scintillation detector (Sigmamotor pump Model T-8 with vernier controlled speed adjustment). The filter (FP) is made of 2-mm bore capillary tubing packed with fine glass wool; this protects the flowmeter (FM) and the anthracene detector from contamination by any debris from gut. Glass to glass fittings are sealed with silicon-rubber gaskets and held together with spring clamps. Pressure connections from plastic tubing to glass balljoints are secured by using polyvinyl chloride connectors with stainless steel fittings (Beckman/Spinco Socket connector, 12 mm No. 312561, Tube Fitting No. 313285 and Swivel Tube Fitting No. 313336). All glass connections from the flowmeter through the scintillation detector are made of 1-mm bore capillary tubing except the necks of the bubble traps (T and T), which have 5-mm bores. The glass stopcocks are pressure cocks with retention fittings. Outlets D and D are used to drain and clean the detector section of the system. Other items in the diagram are as follows: ScS is the scintillation spectrometer; DP is the digital printout system; RRM is the recording ratemeter; and CS is the cannulated segment of small intestine. A, B, and C are stopcocks used to direct the flow of the perfusion solution. E is the exit for the glass shunting system with a ground-glass Luer fitting. It will be noted that the animal is positioned above the reservoir so that the pump lifts the perfusion solution to the animal and through the lumen of the cannulated segment; the solution then drains by gravity back to the reservoir. The scintillation detector (ScD) is kept cooled by circulating 10° water through its heat sink.

directly while the perfusion progresses. The plotted rate count of radioactivity presents an infinite number of analytical points. Concurrent with the recording ratemeter, the level of radioactivity is recorded by digital printout over predetermined time intervals to print as many points of information as seem desirable.

Monitor System for Radioactivity

Prior to perfusion the detector system must be programmed for the detection and recording of the radioactivity level of the perfused material. A sample of material to be measured is instilled into the anthracene-packed detector, and the spectrometer is adjusted to obtain the response desired. The recording ratemeter is set to a slow chart speed (3 in./hr with 5-in. paper width) and the full-scale response is adjusted to maximum deflection, according to the specific activity of the solution being measured. Digital printout is set for 1-min time intervals. Single-channel recording is adequate. However, multichannel recording can be used for mixed isotope studies.

Perfusion

Before perfusion the scintillation detector and the system from stopcock A through cock C and B is filled with water, saline or buffer. The perfusion reservoir is filled with 20 ml of the solution to be perfused through the intestinal lumen of the animal. Cock B is closed and cock A directs the flow to exit E. The tube leading from the glass shunt system to the heat exchanger is disconnected at E, and the solution from the reservoir is pumped through the filter and flowmeter, directed by stopcock A upward to fill the shunt connection to the exit at E; 1 ml is collected and discarded. Cock C is open across from A to the detector; cock B is closed at this time, thus bypassing the detector. Open cock B; direct flow through the detector from A and collect 2 ml from the end of the glass system at E and discard. This fills the detector system with the perfusion solution. Reconnect the plastic tube leading to the heat exchanger from the glass connector (E) and continue pumping the solution through the heating coil to the three-way cock (3/C) and direct the flow back to the reservoir. Continue pumping the recirculating perfusate to establish an initial radioactivity level record. A 1-ml sample is collected for control analyses. While the above proceeds, the fasted animal is anesthetized and cannulated. Cannulation of the upper small intestine is carried out as follows: A midline incision is made to expose the viscera. An in-going cannula is inserted 3.5 to 4 cm distal to the pyloris and secured with a ligature. An out-going cannula is inserted 10 cm distal to the first and secured. The positions of the cannulas are approximated at the time of surgery and measured exactly at the end of the experiment. The surgical area is covered with cotton soaked in saline and kept warm [1].

To start perfusion of the lumen the flow of perfusate is directed (3/C) to the intestinal segment, and "zero time" is set at the time the solution is visualized at the distal cannula. The gut is "washed" by the first milliliter which passes through the system; the "wash" is discarded. The remaining 15 ml are recirculated.

At the end of a monitored perfusion period the perfusate is shunted away from the animal by deflecting the flow at 3/C and stopping the pump. The return tube from 3/C is disconnected and drained to the reservoir. Air is blown through the cannulated intestinal segment from the opened connection at 3/C with a 20-ml syringe (gentle pressure) to clear the intestinal lumen and tubing of any perfusion solution in this portion of the circuit by forcing it into the reservoir. The cock B is closed, and the scintillation detector is bypassed by resetting cock A. The remaining parts of the system are drained, and the volume measured.

The whole system is flushed with 1.0% $NaHCO_3$ and then water. The detector portion of the system is closed off by cocks A and B, and the remainder of the system is flushed with 70% isopropanol, for at least 1 hr. The alcohol is flushed out with water, and this part of the system is drained dry. Upon the preparation of a fresh filter-pack the apparatus is then ready for another study. The detector system from cock A through the detector cell and to cock B is never allowed to become dry or carry any air bubbles. Air trap T protects the detector cell.

Addition of Antagonists

A metabolic antagonist can be added to the system without change in the circulation volume. Place a calculated quantity of the antimetabolite into a dry 2-ml syringe fitted to an 18-gauge needle and polyethylene tubing long enough to reach the contents of the reservoir. At any selected time a small volume of the circulating solution can be drawn up into the syringe to dissolve the antagonist and quickly returned to the reservoir. Should the material be only moderately soluble, flushing of the syringe can be repeated a few times. In this way the perfusion system can be altered in a few seconds. The response of the antagonist as recorded in counts will be delayed only to the extent of the circulating volume and the rate of flow through the system. This will be pointed out under the section on results.

Perfusion Solutions

The amino acids used in these studies were taken up in 0.85% NaCl solution, isotonic to the rat. Tonicity must be maintained to keep the net water exchange in the system under control or minimal since the absorption rate measurements are directly related to concentration of the C^{14}-amino acid in the circulating perfusion solution.

L-leucine, L-isoleucine, and L-aspartic acid were chromatographically pure amino acids obtained from Calbiochem. Radioactive amino acids (u.l. C^{14}) were obtained from the New England Nuclear Corp. Preparations used were checked by chromatography and monitored for radioactivity. Aminooxyacetic acid was obtained from Mann Research Laboratories, Inc.

Animals

Animals used in these studies were adult male rats of Sprague-Dawley origin, maintained on Purina Laboratory Chow until 18 to 20 hr before perfusion. The animals were fasted but allowed water to clear the small intestine of solid dietary debris before perfusion. They were anesthetized by an intra-peritoneal injection of 2% sodium pentobarbital, 50 mg/kg body weight.

RESULTS

The results of an experiment monitored for absorption as it occurs in the perfused intact intestine become apparent within a few minutes after perfusion has begun. The slope of the absorption rate can be visualized in the recorded ratemeter data while the experiment progresses. There is a short delay in recorded data since the material leaving the intestinal lumen must pass through the reservoir before it reaches the scintillation detector. This delay can be seen in the figures which follow.

Absorption of Leucine

A series of experiments were carried out to establish the slope of L-leucine absorption. The studies extended over a range of 0.25 to 5 mM L-leucine. It was found that on oc-casion there was a slight quenching of radioactivity as "seen" by the scintillation detector. This did not affect the slope of absorption and was correctable. Figure 3a presents a tracing of such a record. The initial rate of radioactive count was quenched, however, and the slope which followed was linear and could be extrapolated to a "corrected" initial level of radioactivity. The slope of the absorption rate is also plotted from digital printout data (Fig. 3b), where the means of correction is shown.

Inhibition of Leucine Absorption by Isoleucine

L-leucine was perfused as described, and isoleucine was added to the perfusion system as the experiment progressed. Sufficient isoleucine was added to the reservoir, as described above, to make the perfusion mixture 5 mM isoleucine in the leucine-C^{14} solution which was being perfused with no change in the circulating volume.

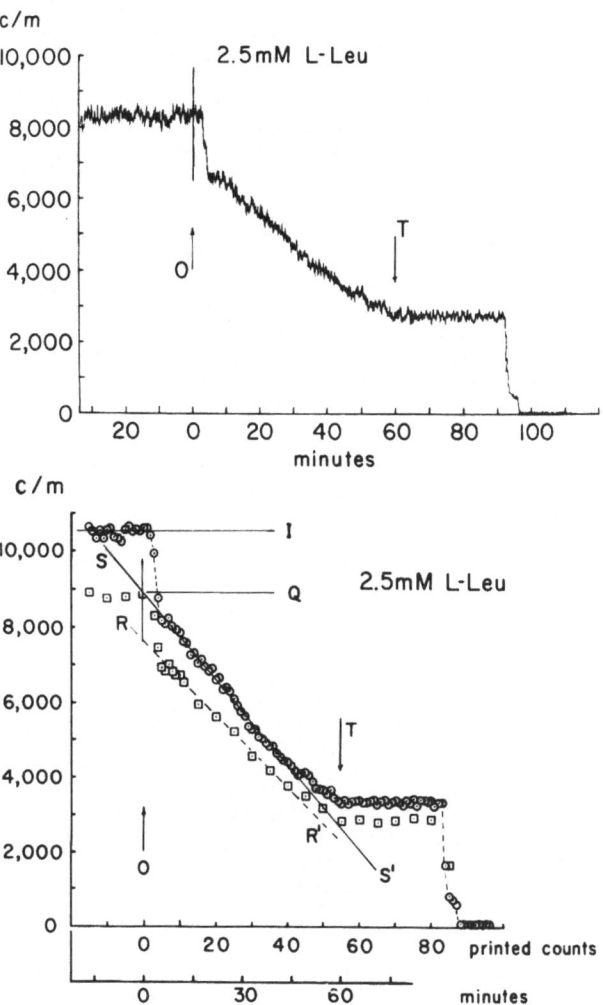

Fig. 3. Quenching Correction. 360-g rat perfused with 2.5 mM L-leucine, 0.028 μc
C^{14}/ml, segment 12 cm long. a. Upper: A tracing of the absorption slope. b.
Lower: Slopes are plotted from digital data. Quenching of radioactivity detection
is apparent in the sudden drop in counts after the perfusion had been started.
Absorption of the amino acid was linear for the first 40 min, at which time intestinal
sloughing was noted with a change in slope of plotted curve, compared with slope
S — S'. Slope S — S' was plotted with 100 to 1000 window setting at full gain and
1095 V. Slope R-- -R' was plotted with a window setting of 100 to 600, full gain.
Quenching from the initial radioactivity level (I) to the "quenched initial level" (Q)
can be plotted to the point where the quenched level at zero time intersects the
slope of absorption. Letter 0 represents "zero time" of perfusion and T the termination of perfusion for these and all subsequent figures.

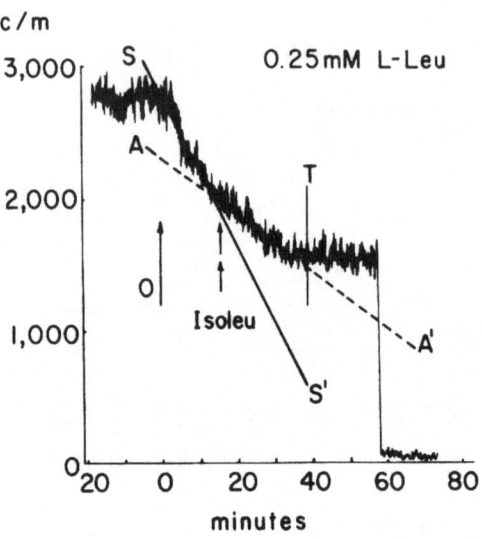

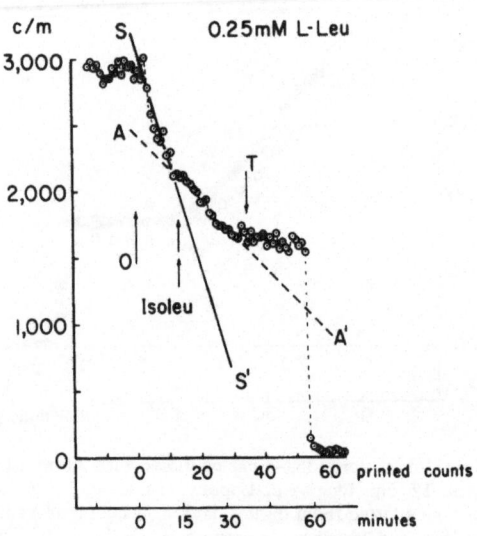

Fig. 4. Antagonism of Leucine Absorption by Isoleucine. 334-g rat perfused with 0.25 mM L-leucine, 0.01 μc C¹⁴/ml, segment 11.5 cm long. Slope S — S' is the absorption slope for L-leucine. Isoleucine to make the system 5 mM isoleucine was added 15 min after perfusion was started. Slope A — — — A' represents the absorption of leucine as affected by its analog isoleucine. a. Upper: Data are presented as a tracing from the recording ratemeter. b. Lower: Slopes are plotted from digital data.

The findings presented in Fig. 4a show a tracing of the radioactivity level of the system when 0.25 mM leucine was perfused, then isoleucine was added at the point indicated in the figure. Figure 4b presents the data as obtained by digital printout after being plotted. Both sets of data present the normal slope of the leucine absorption rate S — S' and the resulting changed slope A — — — A' upon the addition of the second amino acid, demonstrating an antagonism by isoleucine upon the absorption of leucine. These results were obtained in the same preparation with a given intact animal acting as its own control.

A similar experiment is shown in Fig. 5a, where 0.5 mM leucine absorption was antagonized upon the addition of isoleucine. Figure 5b shows these findings after digital data have been plotted.

Competitive Inhibition

Data from a series of experiments were corrected for cannulated segment length. Values for absorption velocities were calculated in terms of μmoles of leucine absorbed per 10 cm intestinal segment per 20-min time interval. These data were graphed as a Lineweaver-Burke plot in order to reveal apparent kinetics and the type of inhibition induced by isoleucine upon leucine absorption (Fig. 6).

The normal reciprocal plot slope (—O—) extends from the concentration level of leucine in blood up to levels above that found in intestinal mucosal juices. The higher levels of leucine lost linearity when diffusive transport was added to active transport. The mucosal concentration of leucine is indicated in the figure by the vertical arrow.

Experiments in which isoleucine had been used showed that the inhibition found upon perfusion followed the pattern characteristic of the competitive type (—Δ—). These points were controlled by the same animal experiment used to establish the normal slope.

Efflux into the Intestinal Lumen

The perfusion system described, monitored for radioactivity, can be applied to studies of secretion into the intestinal

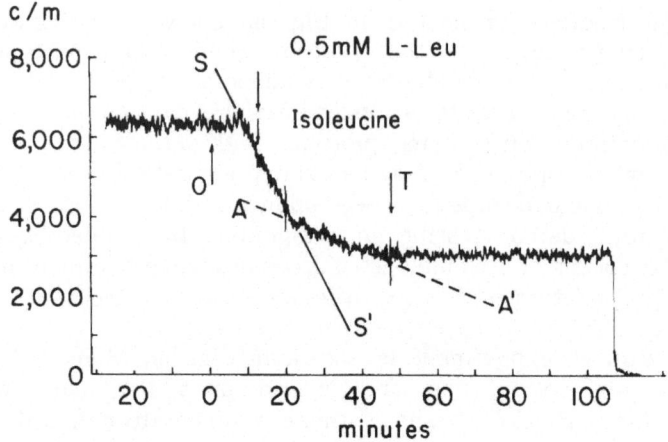

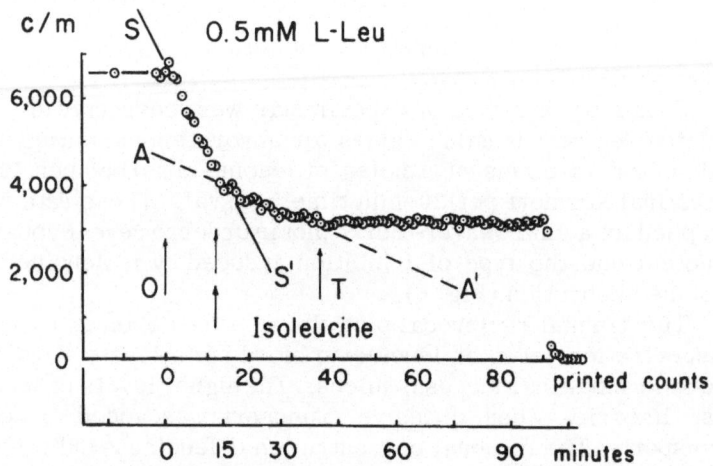

Fig. 5. Antagonism of Leucine Absorption by Isoleucine. Rat perfused with 0.5 mM L-leucine, 0.02 μc C^{14}/ml. Isoleucine was added to the system 12 min after perfusion was started. Slope S — S' represents the absorption of L-leucine and A – – – A' after affected by the presence of isoleucine. a. Upper: A tracing from the recording ratemeter. b. Lower: Slopes are plotted from digital data.

lumen as well as absorption from the lumen. The system is arranged as in the aforementioned experiments but non-radioactive isotonic saline or buffer is perfused.

An example of such a study is presented. The rat was perfused with the salt solution and a baseline in counts per minute was established. The animal was injected intravenously with 0.5 mg L-aspartic acid-C^{14}, 5 μc in a volume of 0.5 ml. The results are shown in Fig. 7. The aspartic acid was detected in the lumenal contents almost immediately.

Since the aspartic acid could have been catabolized in this process as well as reabsorbed by the intestine, the system was stopped and the lumenal contents were analyzed using a Beckman/Spinco Amino Acid Analyzer. The effluent from the column was monitored for radioactivity, which was found to be in the form of aspartic acid. These measurements were made to verify the fact that the radioactivity detected by the monitored perfusion system was indeed the material injected.

DISCUSSION

The perfusion system described has an advantage over a system which requires the removal of samples from the circulating volume for analytical measurements. A continuous plot, related to absorption rate, can be visualized at once with an infinite number of values along the slope. Digital data

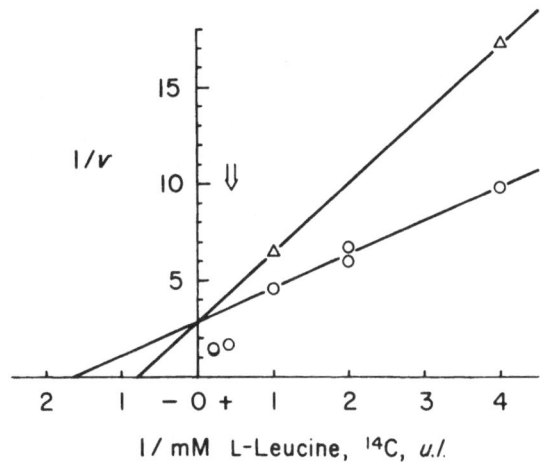

Fig. 6. Competitive Inhibition of Leucine Absorption by Isoleucine. Data were obtained from monitored perfusion experiments. Velocity is defined as micro moles absorbed per 10 cm cannulated intestinal segment per 20 min. The vertical arrow indicates the position for the concentration of leucine in fresh intestinal mucosa of the rat. The slope -0- represents a normal absorption pattern; slope -Δ- results from the presence of 5.0 mM isoleucine, added during the perfusion, as shown in Figs. 4 and 5. All experiments were corrected for segment length.

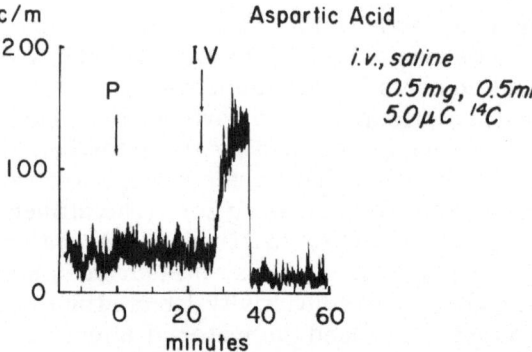

Fig. 7. Aspartic Acid Efflux. This curve reveals the outward transport of L-aspartic acid, C^{14}. The radioactive amino acid was injected via the vena cava at point IV. The appearance of the labeled compound was almost immediate in the lumen. The delay in recorder response is directly related to the volume and flow rate of the perfused system. Starting at point P the animal was perfused with nonradioactive salt solution.

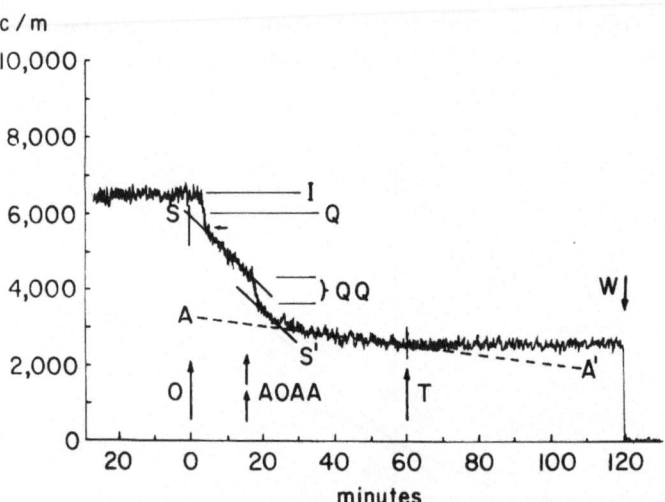

Fig. 8. Quenching by Antagonist. Perfusion of 0.5 mM L-leucine, 0.02 μc C^{14}/ml. The drop from the initial radioactivity level I to the quenched level Q is due to the release of quenching factors from the intestinal lumen. Point Q at an intercept with zero-time 0 permits a corrected initial radioactivity value as can be seen from the slope of line S — S'. The metabolic antagonist aminooxyacetic acid (AOAA) was added (at point AOAA, double arrow) to make the system 0.01 mM AOAA. The drop in the slope S — S' indicated at QQ is quenching due to the AOAA. Slope S — S' represents the normal absorption rate of leucine and slope A – – – A' the inhibited slope by the metabolic antagonist.

printed at 1-min intervals presents as many points of data as there are minutes required for the experiment. With this system any change in linearity of the absorption slope can be noted at once and correlated with other factors pertinent to the animal preparation.

Level of Radioactivity

With low levels of radioactivity the random disintegration rate of C^{14} is quite apparent. Higher levels, 0.02 to 0.03 μc C^{14} per ml of perfusion solution, reveal a smooth statistical pattern. This is apparent in a comparison of digital data plotted in Figs. 3, 4, and 5.

Quenching

The problem of quenching the detectable radioactivity can be corrected since the slopes obtained can be extrapolated. Figure 8 shows two types of quenching. The one has been shown in Fig. 3, a quench related to the biological system or an air bubble carried to the detector. The other quench in count detection (QQ, Fig. 8) is related to the nature of the additive to the system. In this case aminooxyacetic acid quenched the signal but the slope S — S' was continuous until the animal responded to the antimetabolite as is shown by the effected slope A — — — A'.

Biological Problems

Since these studies are carried out with living intact animals, various biological factors have their influences upon the results obtained. The animal is under anesthetic during these experiments, and should respiration be affected, so as to cause erratic breathing and anoxia, the function of the gut would likewise be affected with a resulting progressive change in the slope plotted for adsorption. This can be seen in the experiment cited in Fig. 9. In this experiment the respiration became erratic at point A to become progressively more difficult at points A' and A". After 40 min the animal was moribund.

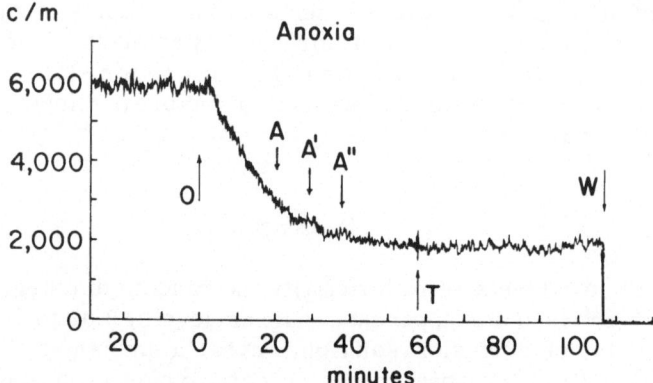

Fig. 9. Anoxia. 220-g rat perfused with 0.5 mM L-leucine, 0.02 μc C^{14}/ml, segment length 9.2 cm. Nonlinearity is related to anoxia. Impaired respiration was noted after 20 min perfusion time; this became progressively more severe, A, A', and A", and was reflected in the progressive change in slope of the absorption rate. The system was washed clean of radioactivity at point W.

These states of anoxia are reflected in the absorption curve. Such an effect upon the absorption rate of the amino acid perfused is in good agreement with in vitro studies which show that the intestinal tissue is demanding of oxygen in carrying out transport processes [8, 9].

Peristaltic activity of the gut or mechanical damage can bring about characteristic patterns of affected absorption rates.

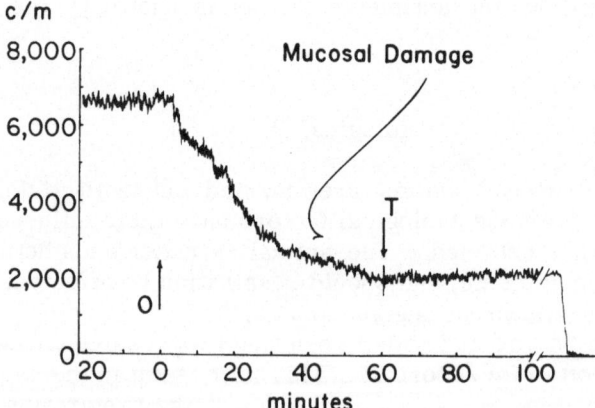

Fig. 10. Mucosal Damage. 241-g rat perfused with 0.5 mM L-leucine, 0.02 μc C^{14}/ml segment length 8.5 cm. The change in linearity of the absorption rate is attributed to mucosal damage, that is, evidence of sloughing of mucosa during perfusion.

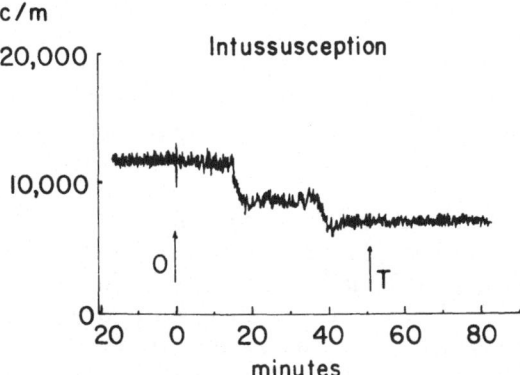

Fig. 11. Intussusception. Blocking of the intestinal lumen by intussusception of the cannulated segment occurred over two intervals, resulting in the stepwise release of altered radioactivity levels into the perfusion system.

On occasion the mucosa will come free of basement membranes in the cannulated segment. This damage to the wall of the lumen will reflect itself in a change in absorption slope, as can be seen in Fig. 10. In the experiment cited there is a definite change in slope concurrent with the mucosal damage.

The peristaltic wave passing down the cannulated segment on occasion can invaginate the gut to produce an intussusception and block the perfusion flow. This leads to a static signal for radioactivity in the detector until the flow is restored, at which time a stepwise pattern would be found, as shown in Fig. 11.

These latter experiments do not present valid findings when one wishes to establish an absorption rate by perfusion. They are situations which do occur and affect the usefulness of the procedure. Problems such as have been presented in Figs. 9 and 10 could not be detected by load-studies, or with difficulty by repeated-sample analyses. Monitored perfusion affords good control and permits the recognition of anoxic and tissue-damage effects.

The addition of antagonists as described permits the same animal to be its own normal control in inhibition or kinetic studies.

SUMMARY

An intestinal perfusion system has been described which has incorporated into it a scintillation detector to provide for continuous monitoring of the radioactivity level of a C^{14}-containing perfusion mixture. Changes in radioactivity levels as affected by intestinal absorption of the lumenal contents have been measured by scintillation spectrometry.

Problems such as anoxia, mucosal damage, and intussusception have been discussed. These are readily recognized with this system.

The system lends itself to kinetic studies of intestinal transport; an example of competitive inhibition of leucine absorption by isoleucine has been presented.

Efflux of C^{14}-containing compounds into the intestinal lumen can be measured. Thus, the system allows for measurements of both absorption from, and efflux into, the intestinal lumen of the intact living animal.

The procedure can be adapted to a variety of biological problems.

REFERENCES

1. Jacobs, F.A., and Luper, M., J. Appl. Physiol. 11:136 (1957).
2. Jacobs, F.A., and Tarnasky, W.G., J. Am. Med. Assoc. 183:765 (1963).
3. Jacobs, F.A., Federation Proc. 24:946 (1965).
4. Jacobs, F.A., Flaa, R.C., Belk, W.F., J. Biol. Chem. 235:3224 (1960).
5. Rapkin, E., and Gibbs, J.A., Nature 194:34 (1962).
6. Elwyn, D.H., Atomlite 37:5 (1964).
7. Jacobs, F.A., and Lang, A.H., Proc. Soc. Exp. Biol. Med. 118:772 (1965).
8. Wilson, T.H., and Wiseman, G., J. Physiol. 123:116 (1954).
9. Wilson, T.H., and Lin, E.C.C., Am. J. Physiol. 199:1030 (1960).

COATED CHARCOAL ASSAY OF VITAMINS, MINERALS, HORMONES, AND THEIR BINDERS

Victor Herbert, Kam-Seng Lau,
and Chester W. Gottlieb
Department of Hematology
The Mount Sinai Hospital
New York, New York

For more than a century, charcoal has been known to be an almost universal adsorbent, capable of adsorbing an almost limitless variety of solids, liquids, or gases from liquid or gaseous solutions [1, 2]. Adsorption by charcoal occurs almost instantaneously, as demonstrated by the rapid decolorization of caramelized sugar solutions when passed through charcoal columns (column chromatography) or shaken with charcoal (batch separation).

It has been known for three decades [3] that charcoal would adsorb vitamin B_{12}; charcoal columns were used in the initial isolation of this vitamin [4]. A number of investigators [5-8] have used charcoal to separate free from bound vitamin B_{12}. These investigators all used 1 ml or more of serum and 100 mg or less of charcoal. To conserve samples for multiple studies, it was our practice to use only 0.1 ml of serum in our researches. Using such serum quantities, we were unable to confirm the work of these earlier investigators since we found that both free and bound vitamin B_{12} were taken up by the charcoal as reported by Kakei and Glass [9]. Reflecting on these seemingly discrepant results, it seemed that the crucial

This work was supported in part by USPHS Grants AM 09564 and AM 09062 and by the Albert A. List, Frederick Machlin, and Anna Ruth Lowenberg Funds.

Dr. Herbert is professor of medicine at the Mount Sinai School of Medicine and holds Career Scientist Award I-435 of the Health Research Council of the City of New York. Dr. Lau is currently head, Clinical Diagnostic Laboratories, Faculty of Medicine, University of Malaya, Kuala Lumpur, Malaysia. Dr. Gottlieb is currently Captain, U.S. Army Medical Corps, Edgewood Arsenal, Edgewood Arsenal, Maryland.

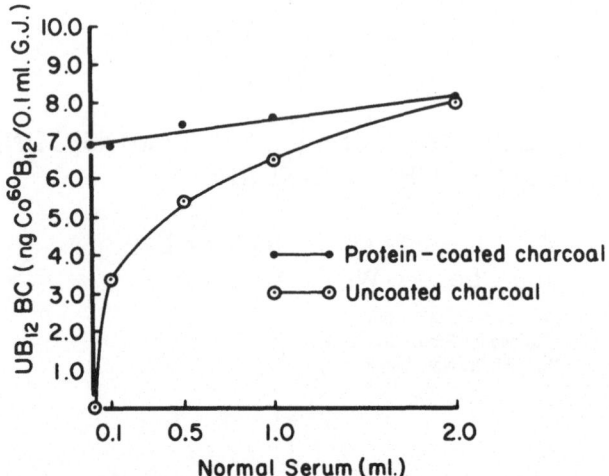

Fig. 1. Measurement of the unsaturated B_{12} binding capacity of 0.1 ml gastric juice using uncoated vs. albumin-coated charcoal. The protein-coating effect of increased amounts of serum is shown.

difference was the quantity of serum used. This led to the speculation that excess serum protein could coat the charcoal to block adsorption of B_{12}-binding protein. It seemed to us that such protein coating could also explain two reports interpreted by their authors to mean that, in the presence of charcoal, intrinsic factor transferred B_{12} to serum [10, 11].

Evaluating this probability, the experiment [12] indicated in Fig. 1 was executed. To a series of test tubes we added 0.1 ml of depepsinized neutralized normal human gastric juice incubated for 30 min with 5 ng of Co^{57}-B_{12}. To five of the test tubes we added quantities of normal human serum varying from 0 to 2 ml, followed by 50 mg of Norit A pharmaceutical grade neutral charcoal in 2 ml of water. Immediately after adding the charcoal the tubes were inverted three times, centrifuged for 10 min at 3000 rpm, and the radioactivity in the supernatant was determined. Five other test tubes were treated identically except that the 50 mg of charcoal used had been previously saturated (coated) with protein by placing it in 2 ml of water containing 20% its own weight (that is, 10 mg) of albumin. As indicated in Fig. 1, albumin-coated charcoal renders the true B_{12}-binding capacity of 0.1 ml of gastric juice even when one adds no serum to it.

The gradual seeming rise in B_{12}-binding capacity of gastric juice indicated on the straight line, as quantities of serum from 0.1 to 2 ml are added, is due to addition of the unsaturated B_{12}-binding capacity of the increasing increments of serum to the fixed unsaturated B_{12}-binding capacity of the 0.1 ml of gastric juice. In striking contrast, when uncoated charcoal is used, less than the true unsaturated B_{12}-binding capacity of 0.1 ml of gastric juice is measured, until one has added 2 ml of serum. This is because 2 ml of serum contains sufficient other protein than B_{12}-binding protein to saturate (coat) the charcoal and thereby prevent the charcoal from adsorbing a significant amount of the B_{12}-binding protein of the gastric juice. However, theoretically there would be a tiny error in determination of unsaturated B_{12}-binding capacity of gastric juice regardless of quantity of serum mixed with it and subsequently added to uncoated charcoal. This would occur because some of the relatively small number of B_{12}-binding protein molecules in the gastric juice would be adsorbed by the uncoated charcoal while it was adsorbing many more of the relatively large number of protein molecules in the serum. Some molecules of gastric juice B_{12}-binding protein would probably always be adsorbed by the charcoal unless the charcoal had previously been coated. This concept is schematically depicted [12] in Fig. 2.

We recognize that this schematic depiction represents a gross oversimplification of the phenomena attending adsorption by coated charcoal. In fact, after more than a century, all the phenomena which produce the adsorption isotherm for any given substance to charcoal have yet to be delineated [1, 2]. The adsorption literature [1, 2] records that the adsorptive capacity of a given charcoal (there are hundreds) for a given agent relates to many variables, including the nature of the charcoal, the agent, and the milieu in which adsorption takes place. These variables include the animal or vegetable origin and internal and external surface areas of the charcoal, its surface pore size and internal channel dimensions, and a variety of surface phenomena related to the particular charcoal being studied. For the agent adsorbed, these phenomena include molecular size and conformation and the effect of the milieu thereon. For example, the conformation of molecules in solution may be sharply changed by small changes of pH and ionic strength of the solvent [13].

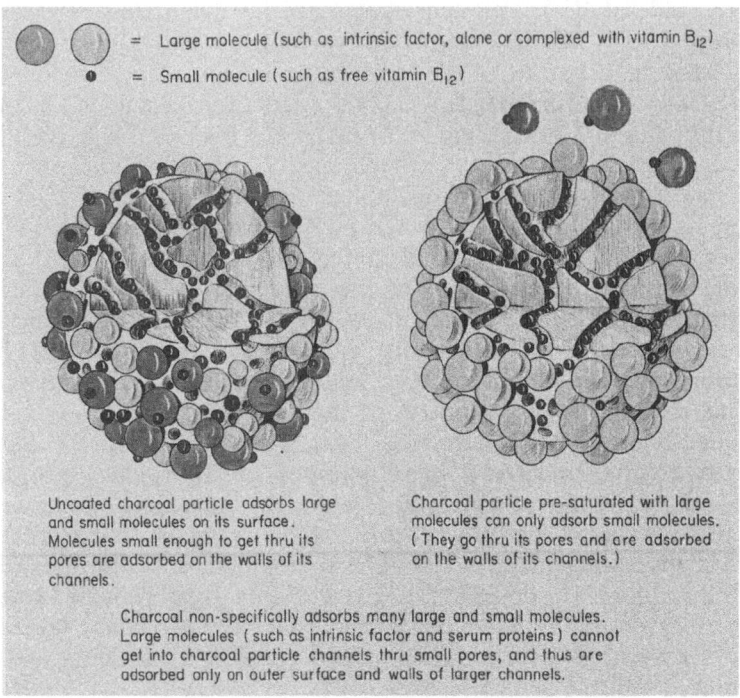

Fig. 2. Schematic depiction of the "instant dialysis" concept, considering the charcoal particle as a "solid microsponge" and the albumin coat as a "molecular sieve" surrounding the sponge.

Oversimplifying, one may state the concept of coated charcoal as the coating of the surface of charcoal particles with a homogenous or heterogenous layer of molecules of a given size and configuration. These molecules should be inert in relation to the molecules to be excluded. This layer constitutes a screen (or molecular sieve) through the pores of which molecules of smaller size and configuration can pass but not molecules of equal or greater size and molecular configuration. The small molecules are adsorbed to the surface of the charcoal particles, between the coating individual larger molecules, as well as being adsorbed along the channels constituting the internal surfaces of the particles. Estimation of molecular weight and configuration of adsorbed molecules is possible by using a series of varying sized coats and determining through which coats the unknown will or will not pass in relation to a series of known molecules.

Putting this simplified concept in a simplified terminology, one may refer to it as "instant dialysis" since it accomplishes within 10 sec the same separation of large from small molecules which can be accomplished by dialysis to exhaustion for 48 to 72 hr. The rapidity of the instant dialysis technique allows its use in purification of protein and other materials without denaturation since materials adsorbed by charcoal may also be desorbed therefrom in an appropriate milieu.

To assay for vitamins, minerals, or hormones, the coated-charcoal concept was married to isotope dilution. Dilution of an unknown with a labeled known quantity of similar material, or "saturation analysis", was used by Landsteiner, who measured an unknown amount of colorless antigen by adding a known amount of yellow antigen, precipitating a known part of the pool of colorless plus colored antigen with a known amount of antibody, and then comparing the amount of color in the supernatant in presence of the colorless unknown with a control containing only the colored antigen and no colorless unknown antigen [14]. The same concept was subsequently applied by Berson and Yalow to assay of hormones, using radioactivity instead of color as the label [15].

Figure 3 illustrates the isotope dilution "saturation analysis" concept. The unknown agent, if bound, is first rendered free (using heat and acid or other appropriate modality which destroys the binder or separates the agent from its binder). A known amount of labeled agent is then added. To the pool of known quantity of labeled agent and unknown quantity of unlabeled agent is then added a fixed quantity of binder for the agent. This quantity of binder must have a total capacity to bind less than the total quantity of unlabeled plus labeled agent present. Appropriately coated charcoal is then used to adsorb all the remaining free agent, which is then precipitated by centrifugation of the charcoal suspension. The bound agent remains in the supernatant, which may then be counted in an appropriate apparatus.

If it is desired to run only one rather than a whole series of controls, it is necessary that the amount of binder added is such that it has a fixed maximal capacity to bind which is exceeded by the amount of the agent present. As Fig. 4 [16] illustrates, the ability of that particular hog intrinsic factor con-

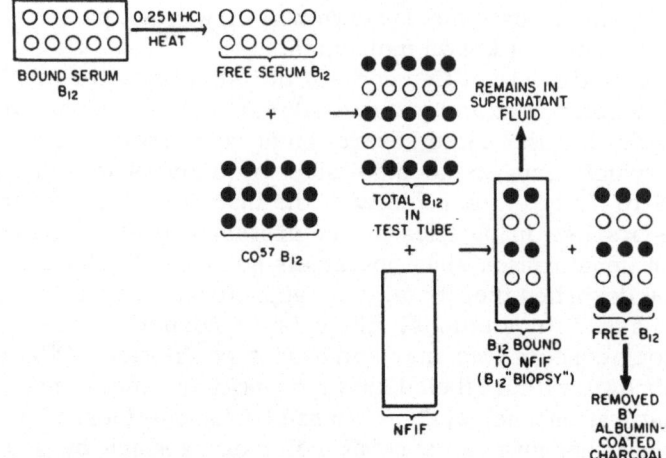

Fig. 3. Schematic representation of serum vitamin B_{12} assay using radioisotope dilution and albumin-coated charcoal. NFIF is the National Formulary Intrinsic Factor.

centrate to bind B_{12} is maximal and fixed only when there are at least 20% more units of B_{12} than there are units of binder. As the number of units of hog intrinsic factor concentrate exceeds the number capable of binding a maximum of 80% of the units of B_{12} present, more than one unit of intrinsic factor is required to bind one of B_{12}. This could be due to association of hog intrinsic factor concentrate molecules with each other. Whatever the cause, it precludes using one control and the formula for a straight line [17, 18]:

$$\text{Weight of unlabeled unknown} = \text{weight of labeled known} \left(\frac{\text{control supernatent counts}}{\text{unknown supernatent counts}} \right) - 1$$

(1)

and makes necessary the use of a series of controls in which the ratio of bound to free agent is plotted on the ordinate as Berson and Yalow have done [15]. To avoid use of such a curve, and to allow use of only one control and equation (1), the binding capacity of any binder under study is determined as indicated in Fig. 4, using a fixed quantity of agent (500 pg of B_{12} in the illustration of Fig. 4) and sequentially increasing the quantity of binder. The quantity of binder selected for use is such as will bind not more than 80% or less than 60% of the fixed quantity of radioactive (or otherwise labeled) agent we propose to add to each sample for assay. Of course, one can use a quantity of binder capable of binding less than 60% of the labeled material, but this would reduce the quantity of radioactivity taken up in the "biopsy" of the pool by the binder.

Another important consideration in coated charcoal assay is recognition that association of agent and binder is rarely instantaneous. The association of B_{12} with hog intrinsic factor concentrate approaches completion at room temperature in approximately 30 min [16]; therefore, 30 min is the time we now select for incubation of B_{12} with intrinsic factor concentrates prior to adding coated charcoal. In our original paper on this subject we stated that 10 sec incubation was enough [12]. This is true only if one meticulously adheres to exactly 10 sec of incubation for each and every sample. It was not until others began having difficulty duplicating our results that we realized the need for a less meticulous procedure, that is, incubation for approximately 30 min, which may vary a few moments from sample to sample without raising problems of duplicability of samples.

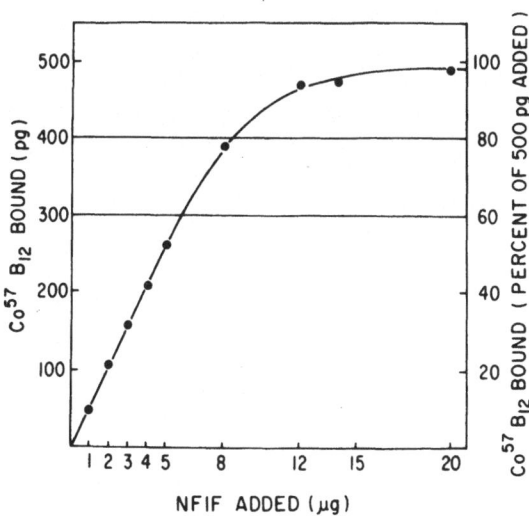

Fig. 4. Percentage of 500 pg $Co^{57}B_{12}$ bound by increasing quantities of National Formulary Intrinsic Factor concentrate (NFIF).

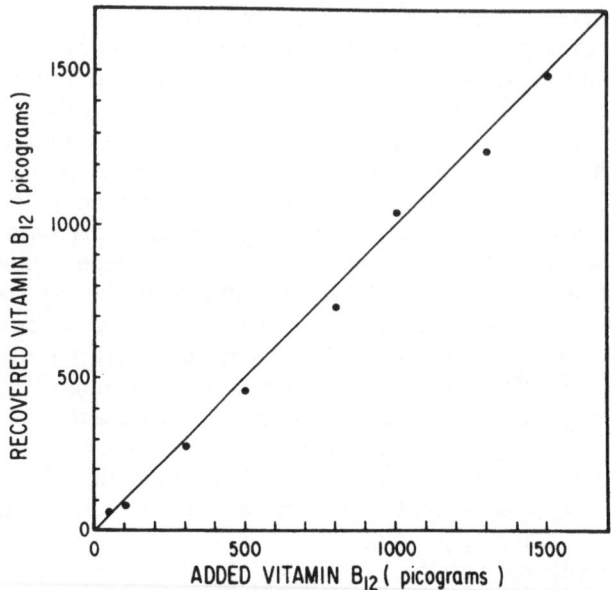

Fig. 5. Recovery of known quantities of crystalline B_{12} added to a B_{12}-deficient serum.

Parenthetically, coated charcoal is a simple and excellent means to delineate the kinetics of association of intrinsic factor concentrate with B_{12} [16] or of any other agent to its binder. The rapidity of association between a fixed quantity of agent and a fixed quantity of its binder under any fixed conditions may be rapidly determined using coated charcoal in the following fashion [16].

One sets up a series of test tubes, each containing the desired fixed quantity of agent under study (that is, 500 pg Co^{57}-B_{12}), in a milieu of desired pH, ionic strength, temperature, and so forth. To each tube one then adds the quantity of binder under study (that is, 7.5 μg of National Formulary Hog Intrinsic Factor Concentrate). To tube 1, one immediately adds appropriately coated charcoal (that is, hemoglobin-coated charcoal). This tube is inverted three times and centrifuged, and the radioactivity of the supernatant is ascertained. In tube 2, agent and binder are allowed to stand for 15 sec prior to adding the coated charcoal; tube 3, for 30 sec; tube 4, for 1 min; and so on for any desired period of incubation prior to adding appropriately coated charcoal. The amount of agent bound after each given period of incubation prior to adding coated charcoal is determined from the supernatant after addition of coated charcoal and centrifugation. The curve thus constructed, with time of incubation on the abscissa and degree of association on the ordinate, delineates the kinetics of association of agent and binder under the established experimental conditions.   For association of iron and transferrin, for example, the curves of association constructed using coated charcoal match curves constructed using much more elaborate biochemical methodology.

Figure 5 [17] shows the excellent recovery of added B_{12} obtained using coated charcoal, one control, and the formula for a straight line—(equation 1).   Figure 6 [17] shows a comparison of 77 coated charcoal assays for serum B_{12} levels with assays obtained using *Euglena gracilis* microbiologic assay. Raven has obtained similar comparisons of coated charcoal with *L. leichmannii* microbiologic assay [19].

Turning to hormone assay, using Dextran 80 (Pharmacia, Inc.) coated charcoal, Fig. 7 [18] shows the percentage of 20 microunits of pork insulin bound by increasing quantities of

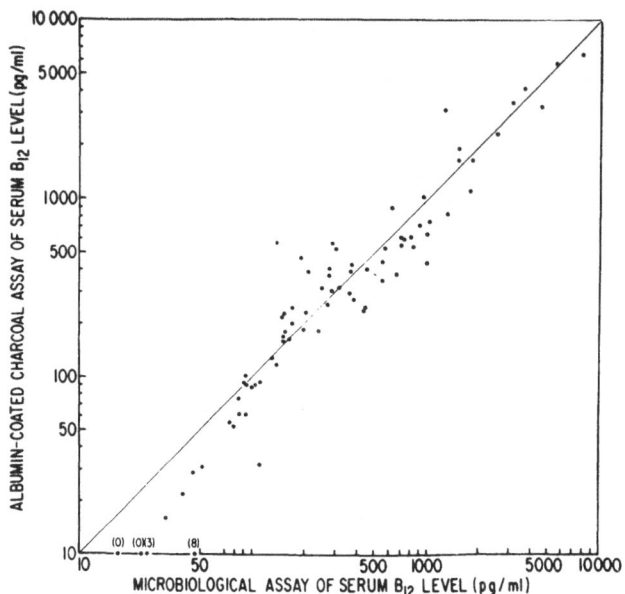

Fig. 6. B_{12} levels of 77 sera as assayed by *Euglena gracilis* and albumin-coated charcoal are compared.   Correlation is sufficiently close to permit similar diagnostic interpretation of each serum B_{12} level.

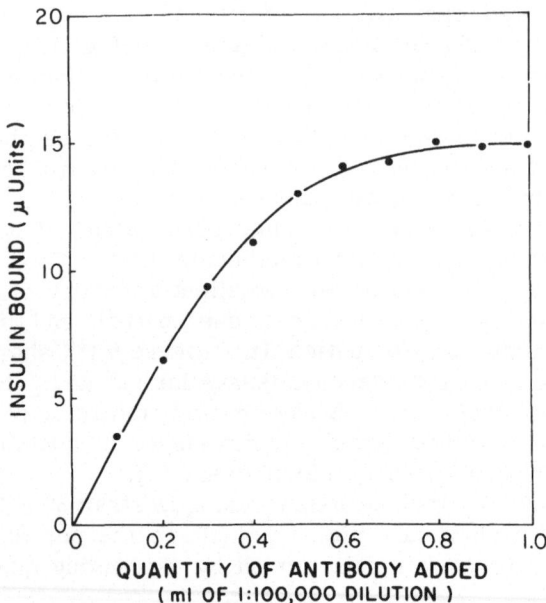

Fig. 7. The insulin-binding capacities of increasing quantities of antibody after incubation with a fixed quantity (20 μU) of insulin are depicted. From this graph the quantity of antibody needed to bind 60 to 80% of the labeled insulin is determined.

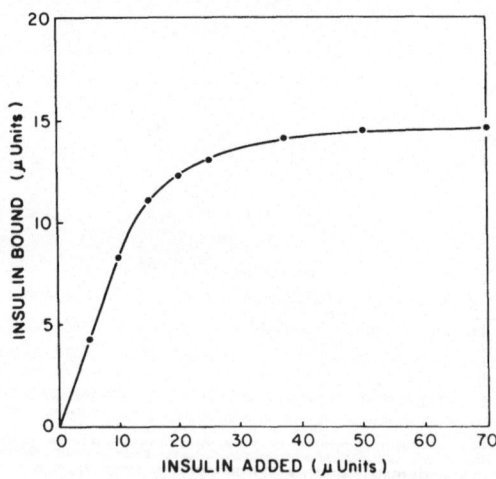

Fig. 8. The insulin-binding capacity of 0.5 ml of a 1:100,000 dilution of guinea pig anti-insulin antibody serum is depicted. The maximum binding capacity, represented by the plateau of the graph, is 14.2 μU of insulin.

a particular batch of guinea pig antibody to pork insulin. The straight-line portion of the curve with lesser quantities of antibody indicates that, at those quantities of antibody, each unit of antibody has a fixed maximal capacity to bind insulin, as supported by Fig. 8 [18]. With greater quantities of antibody there appears to be more than one unit of antibody binding a single unit of insulin, as indicated by the decreasing slope of the curve. This is similar to the binding of B_{12} by hog intrinsic factor concentrate, as indicated by comparison of Fig. 7 [18] with Fig. 4 [16]. If one wishes to assay hormones using the simple formula for a straight line instead of a complicated B/F curve, it is crucial to use a quantity of antibody sufficiently small (in relation to the quantity of radioactive hormone) that it has a fixed capacity to bind hormone. This requires constructing a curve like Figs. 4, 7, or 10 for each new lot of antibody and using regularly a quantity of antibody below the point where the curve ceases to be an ascending straight line (that is, below where the curve starts to level off). If the quantity of antibody used is greater, more than one molecule of hormone will be bound by a single molecule of antibody, and it will become necessary to use a B/F curve rather than the formula for a straight line.

Figure 9 [18] shows an insulin-recovery experiment using coated charcoal assay, one control, and the equation for a straight line—equation (1)—to determine recovery, rather than using a series of controls to plot bound/free ratios. One can use a bound/free control curve, but it is ordinarily unnecessary extra work. A control curve with a series of bound/free points is necessary when a maximal fixed binding capacity for a given quantity of binder is difficult to delineate. Figure 10 [20] shows the binding of a fixed quantity of human growth hormone by increasing quantities of antibody. Although the lesser quantities of antibody do appear to have a relatively fixed maximal capacity to bind human growth hormone, manifested by the lower portion of the curve being straight, it is not as straight as for insulin, and therefore the possibility for error is greater. This may vary with the source of antibody; antibody to insulin and to human growth hormone is commercially available from a number of sources, as is I^{131}-insulin and I^{125}-insulin.

Using a single control point and the equation for a straight line [equation (1)], excellent recoveries of added human growth hormone are obtained (Fig. 11) [20].

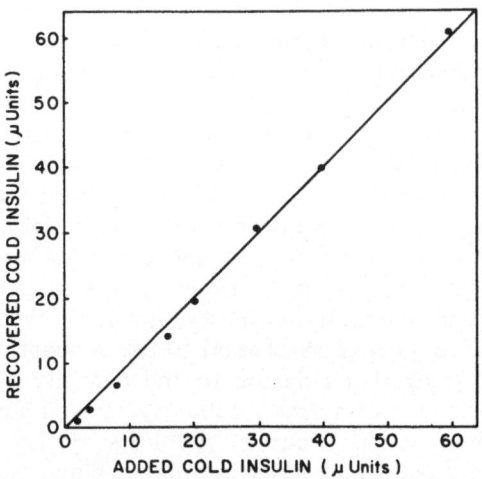

Fig. 9. Recovery of known quantities of crystalline insulin in Dextran-80-coated charcoal insulin assay.

Among the assets of coated charcoal assay of hormones is the fact that the hormone content of an unlimited quantity of serum may be ascertained. This allows assay of very low levels of hormone (by using a large quantity of serum) and also allows use of commercial iodinated hormone of low specific activity, including I^{125}.

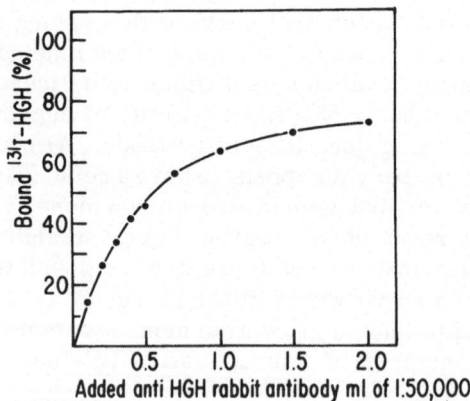

Fig. 10. Binding of a fixed quantity of human growth hormone by increasing quantities of antibody.

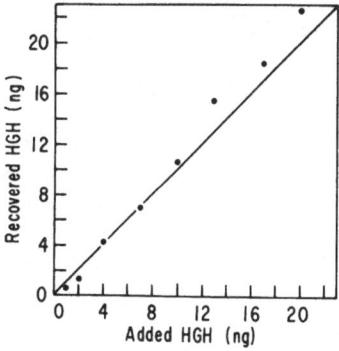

Fig. 11. Recovery of known quantities of human growth hormone in Dextran-250-coated charcoal growth hormone assay.

Figure 12 schematically depicts determination of iron-binding capacity using hemoglobin-coated charcoal [21]. Figure 13 depicts the procedure for determining serum iron using hemoglobin-coated charcoal* [22, 23]. In this latter application, the binder for iron is the endogenous transferrin in the serum. One first determines unsaturated iron binding capacity (UIBC) and then serum iron using equation (2).

$$\mu g \text{ Fe in unknown} =$$

$$\frac{\text{bound Fe}^{59} \text{ in unknown}}{\mu g \text{ Fe}^{59} \text{ added} - \mu g \text{ Fe}^{59} \text{ free}} (\mu g \text{ Fe}^{59} \text{ added} - \mu g \text{ UIBC}) - \mu g \text{ UIBC} \quad (2)$$

Coated charcoal assay may also be used in a test of thyroid function (Fig. 14) [24]. Like erythrocytes or resin, coated charcoal will adsorb radioactive triiodothyronine (T_3) from serum in proportion to the relative affinity of the T_3 for serum-binding proteins and for coated charcoal. The affinity of the serum will be proportional to its endogenous thyroxine-binding capacity which is not yet saturated with endogenous thyroxine. This will be proportional to the thyroid status of the patient.

Figure 15 [25] schematizes the methodology for measuring thyroid function using desorption of T_3 from coated charcoal by serum. Such desorption will be proportional to the quantity of unsaturated thyroxine-binding capacity of the serum and illustrates the fact that a material adsorbed to charcoal

*Hydrodarco B. Norit A pharmaceutical grade neutral will not work as well [23].

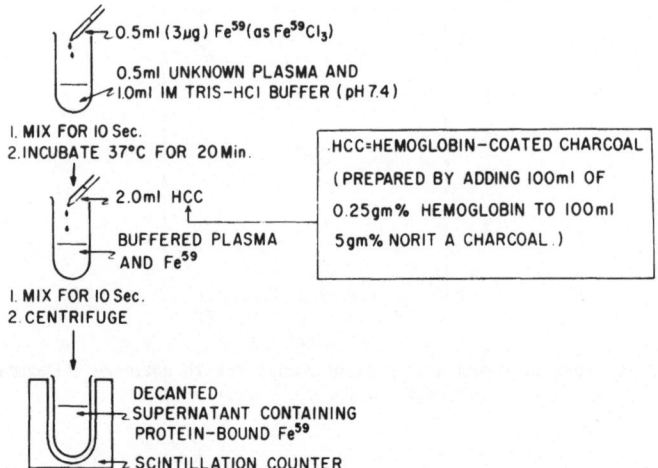

Fig. 12. Schematic depiction of method for determining unsaturated iron-binding capacity of serum using hemoglobin-coated charcoal.

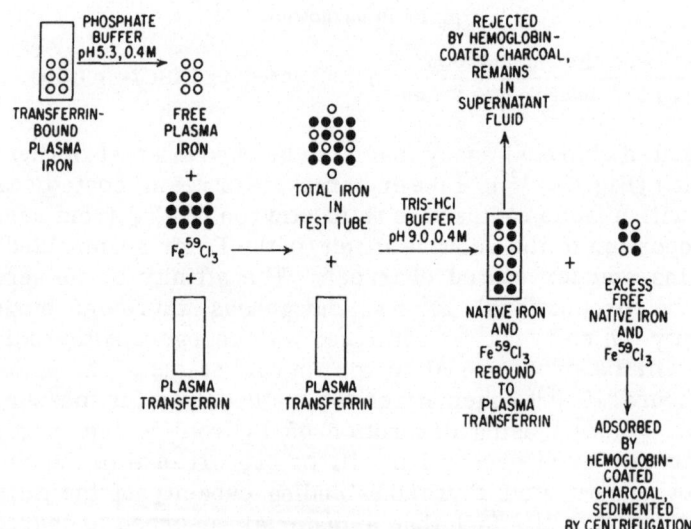

Fig. 13. Schematic representation of serum iron assay using radioisotope dilution, hemoblogin-coated charcoal, and endogenous transferrin (siderophilin) as binder.

may be desorbed therefrom to a variable extent, depending on the nature of the desorbing agent and the incubation milieu and conditions.

The application of coated charcoal to assay of vitamins, minerals, hormones, and their binders, as well as many other materials, seems limitless.   The only requirements are that the material to be assayed must (1) be adsorbable onto charcoal; (2) exist in a labeled (with radioactivity, color, and so forth) form; (3) have a binder, and (4) have a molecular size and configuration (in the milieu used) sufficiently below that of material plus binder so that appropriately coated charcoal will adsorb the free material but exclude the material when complexed with its carrier [26].

Other applications to which coated charcoal has already been extended include assay of chorionic "growth hormone-prolactin" [20], plasma corticosteroids and corticosteroid binding globulin [27], assay of ionized calcium [28], and assay of free thyroxine [29].

It is not necessary that the binder be specific. All that

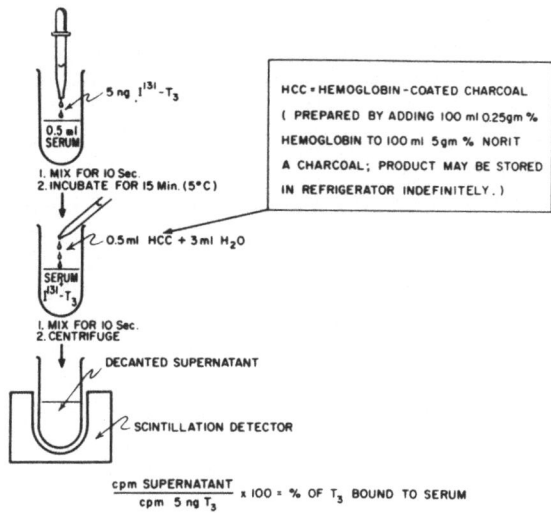

$$\frac{\text{cpm SUPERNATANT}}{\text{cpm 5 ng } T_3} \times 100 = \% \text{ OF } T_3 \text{ BOUND TO SERUM}$$

Fig. 14. Schematic depiction of T_3 test using hemoglobin–coated charcoal. The ratio of 1 part hemoglobin to 20 parts charcoal by weight is used when the hemoglobin is prepared in the laboratory from hemolyzed blood; when commercially purchased crude hemoglobin powder is used, the ratio is 1 part hemoglobin to 10 parts charcoal.

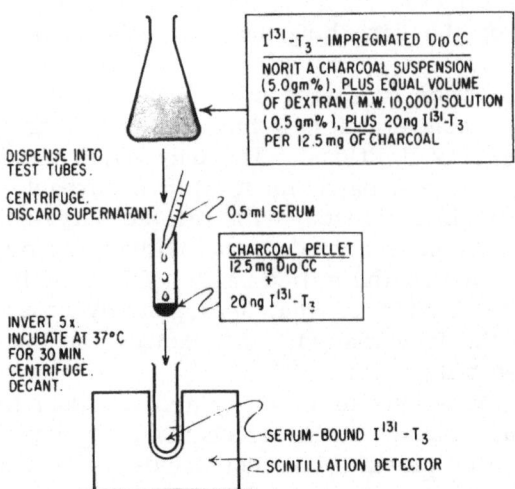

Fig. 15. Schematic depiction of Dextran-10-coated charcoal desorption T_3 test of thyroid function. The net cpm of the unknown samples are compared to the net cpm of a known normal pool and expressed as a ratio.

is necessary is that it have a reproducible binding capacity for the agent to be assayed. For example, although we routinely use hog intrinsic factor concentrate as the binder for B_{12}, we have used saliva, which has a reproducible capacity to bind vitamin B_{12}, and have also used a metabolically dead intrinsic factor concentrate [16]. Of course, it is necessary that the ability of the binder to attach to the unknown is not affected by various possible substances in the milieu or in the unknown and that the unknown not contain a variable amount of binder as well as unknown. Appropriate controls can rule out (or sometimes correct for) these possibilities.

In summary, charcoals (solid microsponges, with small and unchanging void volume) of appropriate particle size, porosity, activity, capacity, and reproducibility may be converted to molecular sieves of tailored pore size by saturation (coating) with an adsorbate of appropriate molecular size and configuration. Such coated charcoal constitutes a system of "instant dialysis" for rapid batch separation of large from small molecules by adsorption of the latter. Coated charcoal can rapidly accomplish many separations previously requiring various forms of adsorption chromatography, dialysis to exhaustion, precipitation by salts, and so forth. Partial purifications of a number of materials may be achieved by adsorption to appropriately coated charcoal and subsequent desorption therefrom.

Molecular weight estimations may be made using a "stable" of coated charcoals, each coated with a dextran of different average molecular weight, from 10,000 to 2 million,* as well as other coating substances such as α, β, and γ globulins, fibrinogen, PVP, Ficoll (molecular weight average 400,000), hemoglobin, and albumin. In such studies, proteins and other materials of known molecular weight can be used either as coating molecules, as calibration proteins, or both. This may prove even simpler and less expensive than the use of gel filtration to estimate molecular weights of proteins [30]. Molecular weight kits of 9 calibration proteins [range cytochrome C (12,400) to pig thyroglobulin (670,000)] are commercially available).†

The marriage of the coated charcoal technic to saturation analysis (isotope dilution) allows rapid, simple, and reproducible assay of trace amounts of a wide variety of vitamins, minerals, hormones, other agents, and their binders in both free and bound forms.

REFERENCES

1. Hassler, J.W., Activated Carbon, rev. ed., Chemical Publishing, New York (1963).
2. Mantell, C.L., Adsorption, 2nd ed, McGraw-Hill, New York, 1951.
3. Laland, P., and Clem, A., Acta Med. Scand. 88:620 (1963).
4. Smith, E.L., Vitamin B_{12}, Methuen, London, 1965.
5. Miller, O.N., Arch. Biochem. Biophys. 68:255 (1957).
6. Rosenthal, H.L., Myers, P.R., and O'Brien, G., in: Heinrich, H.C. (editor), Vitamin B_{12} and Intrinsic Factor, Second European Symposium, Hamburg, 1961, Ferdinand Enke, Stuttgart.
7. Grossowicz, D., Sulitzeanu, D., and Merzbach, D., Proc. Soc. Exp. Biol. Med. 109:604 (1962).
8. Meyer, L.M., Mulzac, C.W., Miller, I.F., and Brennan, B.L., Acta Haematol. 27:229 (1963).
9. Kakei, M., Glass, G.B.J., Proc. Soc. Exp. Biol. Med. 111:270 (1962); see also Gregory, M.E., and Holdsworth, E.S., Biochem. J. 72:549 (1959).
10. Miller, O.N., Arch. Biochem. Biophys. 72:8 (1957).
11. Ardeman, S., and Chanarin, I., Lancet 2:1350 (1963).

*Available as Dextran 10, 20, 40, 80, 110, 150, 250, 500, and 2000 from Pharmacia Fine Chemicals, 800 Centennial Avenue, Piscataway, New Market, New Jersey, 08854.
†Mann Research Laboratories, 136 Liberty Street, New York, New York 10006.

12. Gottlieb, C., Lau, K. S., Wasserman, L. R., and Herbert, V., Blood 25:875 (1965).
13. Craig, L. C., Fisher, J. E., and King, T. P., Biochemistry 4:311 (1965).
14. Landsteiner, K., The Specificity of Serological Reactions, 2nd ed., Harvard University Press, Cambridge, 1945.
15. Yalow, R. S., and Berson, S. A., J. Clin. Invest. 39:1157 (1960).
16. Herbert, V., Gottlieb, C. W., and Lau, K. S., Blood 28:130 (1966).
17. Lau, K. S., Gottlieb, C., Wasserman, L. R., and Herbert, V., Blood 26:202 (1965).
18. Herbert, V., Lau, K. S., Gottlieb, C. W., and Bleicher, S. J., J. Clin. Endocrinol. & Metab. 25:1375 (1965).
19. Raven, J. L., J. Clin. Path., in press.
20. Lau, K. S., Gottlieb, C. W., and Herbert, V., Proc. Soc. Exp. Biol. Med. 123:126 (1966).
21. Herbert, V., Gottlieb, C. W., Lau, K. S., Fisher, M., Gevirtz, N. R., and Wasserman, L. R., J. Lab. Clin. Med., 67:855 (1966).
22. Herbert, V., Fisher, M., Lau, K. S., Gottlieb, C., Gevirtz, N. R., and Wasserman, L. R., Am. J. Clin. Nutr. 16:385 (1965).
23. Herbert, V., Gottlieb, C. W., Lau, K. S., Gevirtz, N. R., and Wasserman, L. R., J. Nucl. Med., in press.
24. Herbert, V., Gottlieb, C. W., Lau, K. S., Gilbert, P., and Silver, S., J. Lab. Clin. Med. 66:814 (1965).
25. Gottlieb, C. W., and Herbert, V., J. Lab. Clin. Med. 68:113 (1966).
26. Herbert, V., Lau, K. S., Gottlieb, C. W., Ann. Intern. Med., 64:1184 (1966).
27. Nugent, C. A., and Mayes, D. M., J. Clin. Endocrinol., 26:1116 (1966).
28. Briscoe, A., and Ragan, C., J. Lab. Clin. Med. 69:350 (1967).
29. Kowal, J., and Sirota, D., in preparation.
30. Andress, P., Nature 196:36 (1962).

INDEX